Food Processing and Preservation Technology

Food Processing and Preservation Technology

BY

Dr. Gboyega Orolugbagbe

2025

Agri Horti Press

New Delhi-110 084

ISBN 978-93-87642-74-4 (HB)

Published By

Agri Horti Press

E-mail: agrihortipress@gmail.com

Printed at : **Replika Press Pvt. Ltd.**

PRINTED IN INDIA

PREFACE

Food Processing and Preservation Technology covers all aspects of the food system from 'farm to fork' such as on-farm production, off-farm processing, distribution and consumers' food selection and consumption, including the effect of different food on their health. The present book provides comprehensive coverage of the processing and preservation techniques of food that include food process engineering, consumption of food processing industries, emerging technology in food preservation, fruit and vegetable processing, preservation of fruits and vegetables, investigation of food borne disease outbreaks, food preservation through pickling and preservation through canning, freezing and drying. This is an ideal text cum reference book for UG and PG students offering courses in food science, food technology and home science. Besides, researchers and professionals in the food processing industry would also find this book extremely rewarding.

I convey my sincere thanks to the publishers, for the meticulous processing of the manuscript, both during editorial and production stages.

Dr. Gboyega Orolugbagbe

Contents

1

Food Process Engineering

Introduction

As an introduction to food process engineering, this book describes the scientific principles on which food processing is based and gives some examples of the application of these principles in several food industries. After understanding some of the basic theory, students should study more detailed information about the individual industries and apply the basic principles to their processes.

For example, after studying heat transfer in this book, the student could seek information on heat transfer in the canning and freezing industries. To supplement the relatively few books on food process engineering, other sources of information are used, for example:

- *Specialist descriptions of particular food industries:* These in general are written from a descriptive point of view and deal only briefly with engineering.
- *Textbooks in chemical and biological process engineering:* These are studies of processing operations but they seldom have any direct reference to food processing. However, the basic unit operations apply equally to all process industries, including the food industry.
- *Engineering handbooks*: These contain considerable data including some information on the properties of food materials.
- *Periodicals:* In these can often be found the most up-to-date information on specialized equipment and processes, and increased basic knowledge of the unit operations.

BASIC PRINCIPLES OF FOOD PROCESS ENGINEERING

The study of process engineering is an attempt to combine all forms of physical processing into a small number of basic operations, which are called unit operations. Food processes may seem bewildering in their diversity, but careful analysis will show that these complicated and differing processes can be broken down into a small number of unit operations. For example, consider heating of which innumerable instances occur in every food industry. There are many reasons for heating and cooling–for example, the baking of bread, the freezing of meat, the tempering of oils. But in process engineering, the prime considerations are firstly, the extent of the heating or cooling that is required and secondly, the conditions under which this must be accomplished. Thus, this physical process qualifies to be called a unit operation. It is called 'heat transfer'.

The essential concept is therefore to divide physical food processes into basic unit operations, each of which stands alone and depends on coherent physical principles. For example, heat transfer is a unit operation and the fundamental physical principle underlying it is that heat energy will be transferred spontaneously from hotter to colder bodies.

Because of the dependence of the unit operation on a physical principle, or a small group of associated principles, quantitative relationships in the form of mathematical equations can be built to describe them. The equations can be used to follow what is happening in the process, and to control and modify the process if required.

Important unit operations in the food industry are fluid flow, heat transfer, drying, evaporation, contact equilibrium processes (which include distillation, extraction, gas absorption, crystallization, and membrane processes), mechanical separations (which include filtration, centrifugation, sedimentation and sieving), size reduction and mixing.

These unit operations, and in particular the basic principles on which they depend, are the subject of this book, rather than the equipment used or the materials being processed. Two very important laws which all unit operations obey are the laws of conservation of mass and energy.

Conservation of Mass and Energy

The law of conservation of mass states that mass can neither be created nor destroyed. Thus in a processing plant, the total mass of material entering the plant must equal the total mass of material leaving the plant, less any accumulation left in the plant. If there is no accumulation, then the simple rule holds that "what goes in must come out". Similarly all material entering a unit operation must in due course leave.

For example, if milk is being fed into a centrifuge to separate it into skim milk and cream, under the law of conservation of mass the total number of kilograms of material (milk) entering the centrifuge per minute must equal the total number of kilograms of material (skim milk and cream) that leave the centrifuge per minute. Similarly, the law of conservation of mass applies to each component in the entering materials. For example, considering the butter fat in the milk entering the centrifuge, the weight of butter fat entering the centrifuge per minute must be equal to the weight of butter fat leaving the centrifuge per minute. A similar relationship will hold for the other components, proteins, milk sugars and so on.

The law of conservation of energy states that energy can neither be created nor destroyed. The total energy in the materials entering the processing plant, plus the energy added in the plant, must equal the total energy leaving the plant.

This is a more complex concept than the conservation of mass, as energy can take various forms such as kinetic energy, potential energy, heat energy, chemical energy, electrical energy and so on.

During processing, some of these forms of energy can be converted from one to another. Mechanical energy in a fluid can be converted through friction into heat energy. Chemical energy in food is converted by the human body into mechanical energy.

For example, consider the pasteurizing process for milk, in which milk is pumped through a heat exchanger and is first heated and then cooled. The energy can be considered either over the whole plant or only as it affects the milk.

For total plant energy, the balance must include: the conversion in the pump of electrical energy to kinetic and heat energy, the kinetic and potential energies of the milk entering and leaving the plant and the various kinds of energy in the heating and cooling sections,as well as the exiting heat, kinetic and potential energies.

To the food technologist, the energies affecting the product are the most important. In the case of the pasteurizer, the energy affecting the product is the heat energy in the milk. Heat energy is added to the milk by the pump and by the hot water passing through the heat exchanger. Cooling water then removes part of the heat energy and some of the heat energy is also lost to the surroundings.

The heat energy leaving in the milk must equal the heat energy in the milk entering the pasteurizer plus or minus any heat added or taken away in the plant. Heat energy leaving in milk = initial heat energy

+ heat energy added by pump

+ heat energy added in heating section

– heat energy taken out in cooling section

– heat energy lost to surroundings.

The law of conservation of energy can also apply to part of a process. For example, considering the heating section of the heat exchanger in the pasteurizer, the heat lost by the hot water must be equal to the sum of the heat gained by the milk and the heat lost from the heat exchanger to its surroundings. From these laws of conservation of mass and energy, a balance sheet for materials and for energy can be drawn up at all times for a unit operation. These are called material balances and energy balances.

Overall View of an Engineering Process

Using a material balance and an energy balance, a food engineering process can be viewed overall or as a series of units. Each unit is a unit operation. The unit operation can be represented by a box as shown in Figure.

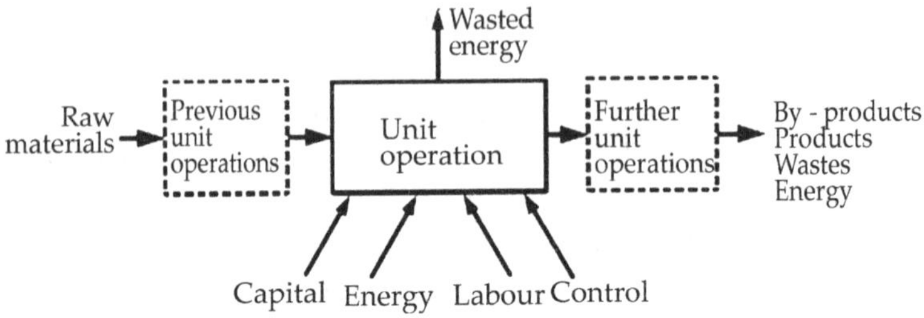

Fig. Unit Operation.

Into the box go the raw materials and energy, out of the box come the desired products, by-products, wastes and energy. The equipment within the box will enable the required changes to be made with as little waste of materials and energy as possible. In other words, the desired products are required to be maximized and the undesired by-products and wastes minimized. Control over the process is exercised by regulating the flow of energy, or of materials, or of both.

Types of Process Situations

Continuous processes

In continuous processes, time also enters into consideration and the balances are related to unit time.

Thus in considering a continuous centrifuge separating whole milk into skim milk and cream, if the material holdup in the centrifuge is constant both in mass and in composition, then the quantities of the components entering and leaving in the different streams in unit time are constant and a mass balance can be written on this basis. Such an analysis assumes that the process is in a steady state, that is flows and quantities held up in vessels do not change with time.

Example: Materials Balance in Continuous Centrifuging of Milk

If 35,000 kg of whole milk containing 4per cent fat is to be separated in a 6 h period into skim milk with 0.45per cent fat and cream with 45per cent fat, what are the flow rates of the two output streams from a continuous centrifuge which accomplishes this separation? Basis 1 hour's flow of whole milk have only recently become commerciably viable. Food preservation using high pressure is a promising technique in food industry as it offers numerous opportunities for developing new foods with extended shelf life, high nutritional value and excellent organoleptic characteristics. High pressure is an alternative to thermal processing. The resistance of microorganisms to pressure varies considerably depending on the pressure range applied, temperature and treatment duration, and type of microorganism. Generally, Gram-positive bacteria are more resistant to pressure than Gram-negative bacteria, moulds and yeasts; the most resistant are bacterial spores. The nature of the food is also important, as it may contain substances which protect the microorganism from high pressure.

This chapter presents results of studies involving the effect of high pressure on survival of some pathogenic bacteria—*Listeria monocytogenes, Aeromonas hydrophila* and *Enterococcus hirae*—in artificially contaminated cooked ham, ripening hard cheese and fruit juices.

The results indicate that in samples of investigated foods the number of these microorganisms decreased proportionally to the pressure used and the duration of treatment, and the effect of these two factors was statistically significant (level of probability, *P* d" 0.001). *Enterococcus hirae* is much more resistant to high pressure treatment than *L. monocytogenes* and *A. hydrophila*. Mathematical methods were applied, for accurate prediction of the effects of high pressure on microorganisms.

The usefulness of high pressure treatment for inactivation of microorganisms and shelf life extention of meat products was also evaluated. The results obtained show that high pressure treatment extends the shelf life of cooked pork ham and raw smoked pork loin up to 8 weeks, ensuring good micro-biological and sensory quality of the products. Food products are an excellent environment for growth of pathogenic microorganisms, which may cause food-borne diseases.

Quality and shelflife of food products depend greatly on the properties of microorganisms contaminating the food. Despite the introduction of food standards obligatory in EU countries, epidemiologists believe that 75 per cent of food-borne diseases are caused by bacteria.

For this reason, the control of microorganisms is an important aspect of food quality and safety. Many methods of food preservation are used for ensuring microbiological safety, among which high pressure processing (HPP) seems a very promising technique for food industry, as it offers numerous opportunities for developing new shelf life stable foods with extended shelf-life, high nutritional value and excellent organoleptic characteristics—minimally processed but safe for consumers. High pressure is an alternative to thermal processing. The resistance of microorganisms to pressure varies considerably depending on the pressure range applied, temperature and treatment duration, and type of microorganism. As a result of technical progress and government support, the first high pressure processed food products appeared in Japan in the early 1990s.

In Europe, high pressure processing (HPP) of foods was rather at the stage of research or pilot production in the last decade. EU legislation included HPP foods in the "novel food" category. EC Novel Food regulation has introduced a statutory pre-market approval system for novel foods across the whole of the European Union.

Recently, rapid progress of HPP towards commercial exploitation has been achieved, but still the process requires close collabouration between researchers, food and equipment manufacturers, as well as proper financial support.

In Poland, research on the application of high pressure for food preservation was initiated in 1992 by the High Pressure Research Centre (now Institute of High Pressure Physics) of the Polish Academy of Sciences in collabouration with others Institutes.

The research programmes, sponsored by the State Committee for Scientific Research and by EU funds, were devoted to high pressure processing of fruit products, fruit and vegetable juices, milk and meat products. This chapter presents some of the results concerning the effect of high pressure on the survival of *Listeria monocytogenes, Aeromonas hydrophila* and *Enterococcus hirae* in artificially contaminated food products. This microorganisms could be a source of food-borne diseases.

Need of high pressure processing

High Pressure Processing (HPP) is a method of food processing wherein the food is subjected to elevated pressures (pressures up to 87,000 pounds per square inch or approximately 6000 atmospheres) with or without the addition of heat to achieve microbial inactivation or to alter the food attributes in order to achieve consumer-desired qualities.

Pressure is effective in inactivating most of the vegetative bacteria at pressures above 400 MPa. HPP retains food quality, maintains natural freshness, and extends microbiological shelf life. HPP can be used to process both liquid and solid (water-containing) foods.

The process is also called as high hydrostatic pressure processing (HHP) and ultra high-pressure processing (UHP) in the literature. High pressure processing causes minimal changes in the fresh characteristics of foods by eliminating thermal degradation. Compared to thermal processing, HPP results in foods with fresher taste, better appearance, texture and nutrition. High pressure processing can be conducted at ambient or refrigerated temperatures, thereby eliminating thermally induced cooked off-flavours. The technology is especially beneficial for heat sensitive products.

Function of HPP

Most processed foods today are heat processed to kill bacteria. Heat oftentimes diminishes the quality of a product. High pressure processing provides an alternative means of killing bacteria which can cause spoilage or food-borne disease without a loss of sensory quality or nutrients. In a typical HPP process, the product is packaged in a flexible container (pouch or plastic bottle) and is loaded into a high pressure chamber filled with a pressure transmitting (hydraulic) fluid.

The hydraulic fluid in the chamber is pressurized with a pump and this pressure is transmitted through the package into the food itself. Pressure is applied for a specific time, usually 3-5 minutes. The processed product is then removed and stored/distributed in the conventional manner. Because of the uniform manner in which the pressure is transmitted (in all directions simultaneously), food retains its shape, even at extreme pressures. And because no heat is needed, the sensory characteristics of the food are retained without compromising microbial safety.

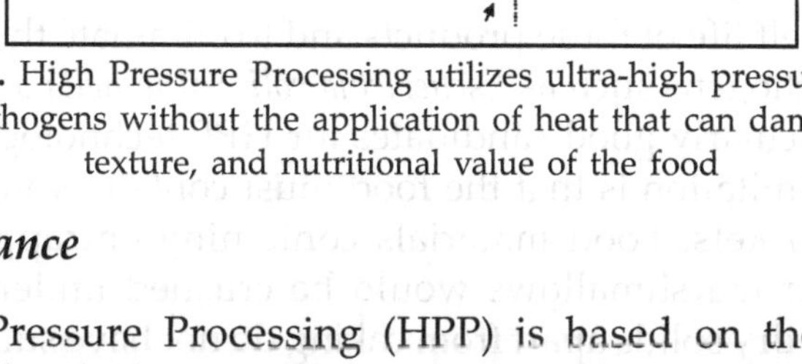

Fig. High Pressure Processing utilizes ultra-high pressures to destroy pathogens without the application of heat that can damage the taste, texture, and nutritional value of the food

HPP at a Glance

- High Pressure Processing (HPP) is based on the Le Chatelicr principle which states that actions that have a net volume increase will be retarded and actions that have a net volume decrease will be enhanced.
- HPP utilizes isostatic or hydrostatic pressure which is equal from every direction.
- During HPP, foods are subjected to pressures up to 100,000 psi. which destroy pathagenic microor-ganisms by interrupting their cellular functions.
- Within a living bacteria cell, many pressure sensitive processes such as protein function, enzyme action, and cellular membrane function

are impacted by high pressure resulting in the inability of the bacteria to survive. Small macromolecules that are responsible for flavour, odour, and nutrition are typically not changed by pressure.

- HPP is gaining in popularity within the food industry because of its capacity to inactivate pathogenic microorganisms with minimal to no heat treatment, resulting in the almost complete retention of nutritional and sensory characteristics of fresh food without sacrificing shelf life.
- One of the unique advantages of HPP is that pressure transmission is instantaneous and uniform, is not controlled by product size and is effective throughout the entirety of the food item.
- As well, HPP offers several advantages over traditional thermal processing including: reduced process times; minimal heat damage problems; retention of freshness, flavour, texture, and colour; and no vitamin C loss.

Like any other processing method, HPP cannot be universally applied for processing all types of foods. At the moment, HPP is being used in the United States, Europe and Japan on a select variety of high value foods either to extend shelf life or to improve food safety. Some products that are being commercially produced using HPP are cooked ready-to-eat meats, avocado products (guacamole), tomato salsa, applesauce, orange juice and oysters.

HPP cannot yet be used to make shelf-stable versions of low-acid products such as vegetables, milk or soups because of the inability of this process to destroy spores. However, it can be used to extend the refrigerated shelf life of these products and to eliminate the risk of various food-borne pathogens such as *Escherichia coli, Salmonella* and *Listeria*. Acid foods are particularly good candidates for HPP technology.

Another limitation is that the food must contain water and not have internal air pockets. Food materials containing entrapped air such as strawberries or marshmallows would be crushed under high pressure treatment, and dry solids apart from caking do not have sufficient moisture to make HPP effective for microbial destruction.

During HPP processing, pressure is uniformly applied around and throughout the food product. For example, a grape placed between fingers can be easily squeezed and broken; this is because the pressure is not applied evenly from all sides simultaneously. On the other hand, if the same grape is squeezed from all sides simultaneously, it will not be crushed. This can be demonstrated by placing a grape inside soda bottle filled with water. By squeezing the bottle, you pressurize the water inside as well as the grape. Yet the grape is not damaged, no matter how hard you squeeze. In the same way, foods processed by high pressure will not be damaged by the applied pressure.

Shelf life of HPP processed product

In general, HPP can provide shelf lives similar to thermal pasteurization. Pressure pasteurization kills vegetative bacteria and, unless the product is acidic, it requires refrigerated storage. For foods where thermal pasteurization is not an option (due to flavour, texture or colour changes) HPP can extend the shelf-life by 2-3 fold over a non-pasteurized counterpart, and improve food safety. As commercial products are developed, shelf life can be established based on microbiological and sensory testing.

High pressure processed products are commercially available in the United States, European and Japanese retail markets. Examples of high-pressure processed products commercially available in the US include fruit smoothies, guacamole, ready meals with meat and vegetables, oysters, ham, chicken strips, fruit juices, and salsa. Low acid shelf-stable products such as soups are not commercially available yet because of the limitations in killing spores with HPP. This is a topic of current research.

It has been generally known that high pressure has very little effect on low molecular weight compounds such as flavour compounds, vitamins and pigments compared to thermal processes. Accordingly, the quality of HPP pasteurized food is very similar to that of fresh food products and the quality degradation is influenced more by subsequent storage and distribution rather than the pressure treatment. Pressure also provides unique opportunity to create and control novel food textures in protein based foods.

In some cases, pressure can be used to form protein gels and increase viscosity without using heat. HPP products currently marketed worldwide are primarily distributed refrigerated. In some cases this is necessary for safety (to prevent the growth of spores in low acid foods). For acid foods, refrigeration is not a necessity for microbial stability, but is employed to preserve flavour quality for extended periods of time.

High-pressure food processing of rice and starch foods

The use of pressure (P) in addition to temperature (T) has been accepted in the field of food science and technology over the past 15 years, and research and development are now under way in the food industry, universities, and government institutes. Several commercial products are now on the market that are prepared using high-pressure techniques. This section describes the principle of high-pressure treatment and the effects of high pressure on foods, with emphasis on the high-pressure effects on starches. Finally, recent successes of the Echigo Seika Company in the application of high-pressure techniques to rice and rice products are discussed.

Principle and Method

High pressure means high hydrostatic pressure generated by the compression of water. A pressure of 100 MPa or higher (usually lower than 1,000 MPa) is used under temperatures below 100 °C. A unit of pressure is expressed in kg cm^{-2}, bar, or pounds in^{-2}, but now the international Pascal unit is commonly used. Therefore, 1,000 bar or 1,000 kg cm^{-2} almost equals 100 MPa. Food, which is contained in a plastic bag and sealed by carefully removing air, is placed in a pressure vessel filled with water and high pressure is then generated in the vessel by a pressure pump.

Pressurization of Food

An egg is not crushed by compression at 600 MPa, but the egg white and yolk are coagulated. The colour of egg yolk of a pressurized egg is naturally yellow, whereas a boiled egg changes to a faded yellow. The colour difference is attributed to changes associated with pressurization: pressure affects only noncovalent bonding and coagulates proteins without splitting covalent bonds, thus keeping the colour and smell intact. High pressure induces protein denaturation in the same way as high temperature. Meat protein is also denatured by pressure treatment at 400 MPa, preserving the properties of raw meats. An example of prawns or shrimp is interesting: although a boiled prawn turns red and the meat coagulates, the appearance of a pressurized prawn is the same as that of raw shrimp, but the meat coagulates after pressurization at 400 MPa for 10 min. High pressure also has an effect on starches: a thick suspension of rice starch forms a ball by standing against its own weight after pressurization at 700 MPa, indicating that starches are gelatinized by the pressure treatment.

Versatile Utility of High Pressure

A high-pressure treatment generally coagulates protein, thereby inactivating the enzymes, gelatinizing the starches, and killing microorganisms. Thus, the use of high pressure promises to be a versatile process in food science and technology. Pressure produces a new texture in meats and starch-based foods, while keeping the original nutrients, flavour, and colour.

High-Pressure Effects on Pure Starches

High-pressure-induced Gelatinization of Starch

Effect of Pressure on Amylase Digestibility of Starches: The starches of potato, maize, and wheat are gelatinized by pressure treatment at warm conditions of 45–50°C. The pressurization produces unique properties that are different from those of heat-gelatinization: heat-treatment destroys starch granules, resulting in a transparent solution, but a pressure-treatment swells the granules while maintaining the granular structure.

Nevertheless, amylolytic enzymes such as α-, β-, and glc-amylases digest the pressurized starches well, being similar to the phenomenon in which the pressure-treatment of proteins increases protease digestibility. The pressure-induced gelatinization of starches exhibits a sigmoid curve, suggesting that a two-state transition is involved, as in heat-induced gelatinization.

Effect of Pressurization Time on Amylase Digestibility of Starches: To attain full amylase digestibility of starches, pressurization under warm conditions for 2 to 6 h is necessary. Interestingly, pressurization of starches for a longer time makes amylase digestion difficult: the amylase digestibility of starches decreased by 20–50 per cent after pressurization for 17 h compared with the maximum digestibility obtained after pressurization for 2 to 6 h. These observations suggest that pressure induces gelatinization of starches, similar to heating, but prolonged pressurization produces a new stable structure of starches, which is not susceptible to attack by amylase. *Birefringence of Starches after Pressurization.*

The birefringence of starches is lost as increasing pressure is applied. Wheat starch is sensitive to pressure and birefringence is lost at 200 MPa. In the pressurization of starches, the number of granules exhibiting complete birefringence decreases without increasing the incomplete birefringent granules. This loss of birefringence shows that the crystalline structure is destroyed by high pressure as well as by high temperature and that the high-pressure-induced gelatinization of starches follows a two-state transition without an intermediate state of destruction.

Physical Properties of Pressurized Starches

When a 50 per cent water suspension of starches is pressurized at 100 to 500 MPa and at 45°C for 1 h and air-dried, followed by analyses of their physical properties, the results are as follows.

According to X-ray crystal analysis, the crystalline structure of potato, waxy maize, and maize starches decreases with an increase in pressure, although the crystalline structure of potato starch does not change up to 500 MPa. The decrease in the high-pressure-induced crystalline structure is parallel with the increase in amylase susceptibility. Amylograms show that the transition temperature of pressurized starches is elevated, thus decreasing the viscosity.

Interestingly, differential-scanning calorimetry (DSC) of pressurized starches shows that the peak temperature of the DSC patterns increases while the peak area decreases, indicating that some pressure-induced structural state of the pressurized starches is perturbed by low energy at higher temperatures.

Summary of Pressure-induced Changes in Starches

The structure of pressurized starches changes accompany-ing the increase in amylase digestibility, a loss of birefringence, and a loss of crystallinity. These unique structural properties, which are somewhat different from those of heat-gelatinized starches, should be analysed in detail for effective applications of high pressure to starch and related foods.

Rice-based Foods Produced by High-pressure Processing

Dr. Akira Yamazaki, of Echigo Seika Company Ltd., studied the properties of pressurized rice in detail to introduce the use of high pressure to the food industry. He equipped several large highpressure machines in his own factory and succeeded in sending rice and rice products to the market by introducing the high-pressure technique to the manufacturing process of rice products, cooked rice (*gohan*), rice crackers (*osembei*), and rice cakes (*omochi*).

High-pressure Effects on Rice Grains

Traditionally cooked-rice grains change their shape and develop some cracks, but rice grains after high-pressure pretreatment followed by heat-cooking swell, and their original shape is maintained without cracks.

High-pressure Cooked Rice for Microwave Ovens

Bread is purchased in the store and toasted just before eating. This is a typical life style, especially on a busy morning. However, 45 min are required to cook rice. Cooked rice tastes best and has its best texture just after steaming. Warming of cold cooked rice makes the taste worse. In general, heating a starch-based food twice leads to an unpalatable food. In history, instant-rice is a dream of Japanese consumers.

Dr. Yamazaki succeeded in producing oven-cooked rice with a good taste and texture for consumers, although the production scale is only enough to fulfill the requirement of a small city. He also succeeded in producing instant rice containing miscellaneous cereals.

These instant cooked-rice cereals exhibited good taste and good texture by a 3-min heating in a microwave oven.When high-pressure pretreated and successively heatcooked rice is compared with traditionally cooked rice, its properties are as follows: first, it is more gelatinized; second, it is more slowly retrograded; and third, it is gelatinized to a greater extent by heating just before serving. Japanese consumers who are particularly sensitive to rice accept these properties.

High-pressure Rice Cakes

During New Year days, the Japanese public is freed from the kitchen and enjoys the New Year cerebrations, eating rice cakes (*omochi*) every day, which are a preserved food. The company introduced the high-pressure technique for producing the traditional rice cake and a special kind of *omochi*, which contains herbs with a natural colour, smell, and taste, and this is now available on the market.

Merits of High-pressure Food Processing

High-pressure processing is useful not only for producing high-quality food, but also for improving the manufacturing process. The production time for rice crackers is shortened by introducing a high-pressure technique into the traditional processing system and the total energy cost is decreased by 10 per cent and the labour force by 56 per cent. The use of high pressure for food processing and cooking, in addition to heating and cooling, is now in our hands. Two factors, T and P, are useful for the manufacturing of good foods, including starch-based foods. Although T and P are used independently, their combined use is also important for the optimal use of high pressure. For example, heat-tolerant bacterial spores are inactivated by pressurization under elevated temperature.

Industrial application of the process faces engineering problems

So far only applications on a pilot plant scale are reported in the literature. For further developments on a larger scale, theoretical and practical problems should be solved. The industrial application of the process faces engineering problems related to the movement of great volumes of concentrated sugar solutions and to equipment for continuous operations. The use of highly concentrated sugar solutions creates two major problems. The syrup's viscosity is so great that agitation is necessary in order to decrease the resistance to the mass transfer on the solution side. The difference in density between the solution (about 1.3 kg/litre) and fruit and vegetables (about 0.8 kg/litre), makes the product float. Another important aspect, so far not investigated, is the microbiological safety of the process, which should be studied thoroughly before further industrial development.

Osmoappertisation in the Processing of Apricots

In order to obtain an alternative to the canned fruit preserves and to maintain a high quality of the fruits, a research has been carried out on the osmoappertisation of apricots, a "combined" technique that consists in the appertisation of the osmodehydrated apricots. This technique could contributes also to the reduction of energy consumption, limits the cost of production and combines "convenience" (ready-to-eat, medium shelf-life) with many market outlets (retail, catering, bakery, confectionery, semi-finished products).

Apricot Processing

- *Fresh apricot puree:* After washing, cutting and removal of stones, apricot halves are dipped in 2per cent solution of sodium or potassium metabisulphite for 10 minutes. After draining, the resulting material is passed through a 0.045-in. screen pulper - finisher to produce a fresh apricot puree. The fresh apricot puree obtained in this way could be further processed in different semiprocessed (*i.e.* chemically or otherwise preserved products) or finished fruit products (fruit leathers, fruit bars, jams, etc.).
- *Concentrated apricot pulp:* Fresh apricot halves could also be steam blanched for 5 min., passed through a 0.045-in pulper - finisher and transformed in a purée with about 14 Brix depending on the fruit quality. This purée may be concentrated in steam jacketed kettles up to 20 Brix or in other adequate equipment (*e.g.* a stirred vacuum evaporator) up to 28°Brix. As for fresh apricot purée, the concentrate may be further processed in various semiprocessed or finished fruit products as mentioned above and as will be described below.
- Dried apricot leather
  - From fresh fruit purée by drum drying. The fresh apricot purée at about 14 Brix could be dried to 12per cent moisture apricot sheets, using a double-drum dryer operating at 132 degrees C with a drum clearance of 0.008 in and speed of 45 sec per revolution.
  - From fruit concentrate by drum drying. The concentrate could also be dried to 12per cent moisture fruit sheets by the same process as described above.
  - From fresh fruit purée or from apricot concentrate by sun/solar drying or by dehydration.
- *Trays:* For sun/solar drying or dehydration of fruit pulp, the trays must have a solid base in order to retain the liquid contents. They may be made of metal, timber or plastic. Stainless steel or plastic trays are most suitable because they are unaffected by acid fruit

pulp; they are, however, expensive. A metal tray could be 75 x 50 cm in size and with side 5 cm high. The trays must keep level during drying; if the tray is not level the pulp will run to the lowest point, giving a layer of irregular depth which will dry unevenly. Any tray which is not made of stainless steel or plastic must be covered inside with a sheet of heavy gauge plastic film to protect the pulp from chemical or bacteriological contamination. Standard sun/solar trays as described can be used by covering them inside with a sheet of plastic film to create a solid base.

- *Preparation before drying/dehydration*: Fresh apricot purée can be directly used for next processing steps. Fruit concentrate needs to be added to potassium metabisulphite to obtain a 0.3per cent concentration of SO2 in the material.
- *Drying/dehydration*: The apricot/fruit purée or concentrate is poured into the trays to a depth of about 1.5 cm. When stainless steel or plastic trays are used they should be coated with a thin layer of glycerine to prevent sticking.

The pulp is then sun/solar dried or tunnel/cabinet dehydrated; moisture content in the dried product should not exceed 14per cent and the SO_2 content should not be less than 1500 pp. The dried product is wrapped in cellophane to prevent sticking, then put inside polythene bags and stored at best in tight fitting tins and sealed to prevent moisture transfer.From fresh fruit purée or from apricot concentrate, with sugar addition, and then processed by sun/solar drying or by dehydration. In some countries preference is for finished products with added sugar; and this is also interesting from a point of view of energy consumption (concentration is partially achieved by sugar dry matter) and of shelf life. The overall content in SO_2 could also be reduced as sugar is a preservation agent, the product will be close to a fruit "paste".

Reconstitution Test for Dried/Dehydrated Products

In reconstitution water is added to the product which is restored to a condition similar to that when it was fresh. This enables the food product to be cooked as if the person was using fresh fruit or vegetable. All vegetables are cooked but many of the dried fruits can be used for eating after they have been soaked in water. The following reconstitution test is used to find out the quality of the dried product. Reconstitution test

- Weigh out a sample of 35 grams from the bulked and packed final product of the previous day's production.
- Put the sample into a small container (beaker) and add 275 ml of cold water (and 3.5 g salt).

- Cover the container (with a watch-glass) and bring the water to the boil.
- Boil GENTLY for 30 minutes.
- Turn out the sample onto a white dish.
- At least two people should then examine the sample for palatability, toughness, flavour and presence or absence of bad flavours. The testers should record their results independently.
- The liquid left in the container should be examined for traces of sand/soil and other foreign matter.

This test can be used also to examine dried products after they have been stored for some time. Evaluation of rehydration ratio may be performed according to the following calculations. Rehydration ratio. If weight of the dried sample is 10 g (Wd) and the weight of the sample after rehydration is 60 g (Wr), rehydration ratio is:

$$\frac{Wr}{Wd}=\frac{60}{10}=\frac{6}{1},6 \text{ to } 1$$

Rehydration coefficient. The weight of rehydrated sample is 60 g (Wr); the weight of dried sample is 10 g (Wd) and its moisture is 5per cent (Wu); raw material before drying had 87per cent water (A); rehydration coefficient is:

$$\frac{Wr}{\dfrac{Wd - Wu \times 100}{100 - A}}$$

$$=\frac{60\times(100-87)}{10-[10\times0.05]}=\frac{780}{9.5}=52.1.$$

A simpler test for eating quality can be carried out without weighing and measuring. The material is placed in a cooking pot with water (and a little salt). The pot is then covered and boiled as described above. Except for a few products which are eaten in the dry state, most dried fruit and all dried vegetables are prepared by soaking and cooking. Often this preparation is carried out incorrectly and dried products get a bad reputation. Good quality dried products, after cooking and if properly treated should be similar to cooked fresh produce. In order to get good results, the following methods are recommended:

Quick Method

Cold water, ten times the weight of the dry product, is added to the dried product. The container is covered, brought to the boil and simmered GENTLY until the product is tender. The cooking time may be 15 to 45 minutes after the boiling point has been reached.

Slow Method

This gives better results than the quick method. Cold water is added to the dry food and is left to soak for 1 to 2 hours before cooking. The product is then cooked in the same water as that in which it was soaked. The actual cooking time will probably be shorter than that for the quick method. Other points to remember are:

- If too much water is added the cooked product will have little flavour. However, if too little water is added the product may dry and burn. This can be avoided by adding small quantities of water during cooking;
- Always cook with a lid on the container;
- Salt, if required, should be added when the cooking is almost complete;
- Partly used packages of dry products should be reclosed tightly or kept in containers with good fitting lids.

2

Consumption of Food Processing Industries

Introduction

One of the most important challenges facing the country is providing remunerative prices to farmers for their produce without incurring the additional burden of subsidies. This challenge could be addressed if the processing level and value addition of the agricultural raw produce can be enhanced to meet the growing demand for processed foods. Food processing has an important role to play in linking Indian agriculture to consumers in the domestic and international markets.

The role of food processing becomes critical since food production is targeted to double in the next 10 years. With low farmer price realizations and significant wastage in the food supply chain even with the current level of production, only processing of food can secure farmer incomes against a slump in prices as well as reduce wastages. Further, a vibrant food processing industry can be a catalyst for crop diversification. The impact of increased economic growth in agribusiness through food processing can play a significant role in reducing rural poverty and increasing rural incomes.

Food processing leads to significant employment generation- not only directly but also indirectly across the supply chain in production and procurement of raw materials and distribution of food products to consumers.

Conditions and Trends of Consumption

In feeding stuffs very great quantities of fats disappear but, with a few unimportant exceptions, they always do so as a constituent of some element of the feed and not as such. Thus the amounts of fat that disappear with grain and legumes fed to animals are truly enormous. In food, as in feeding stuffs, very large quantities of fats are consumed incidentally in the ingestion of meats, fish, poultry, milk, and other animal products and in the form of fats and oils naturally contained in cereals, legumes, fruits, vegetables, and other plant products. However, in the human dietary fats and oils are also consumed practically as such, for example as butter, salad oils, and cooking fats.

Quantitative Data Unsatisfactory

Not only because fats and oils disappear both in the arts and in the dietary in a vast variety of ways, but for other important reasons, the quantitative study of the consumption of fats and oils encounters great difficulties. In the first place, the very concept of consumption is elusive. Presumably one should exclude, in spite of its potential significance, the fat content of large numbers of meat animals that perish from natural causes and from which no attempt is made to recover the fats.

Commercial consumption is a significant concept, but it excludes a large amount of fats in products that do not enter into trade, and a large amount that are incidental components of meats and other products that do enter into trade, for example, tankage and garbage. Wastes in the process of recovery are considerable; hence the fat content of the product treated may be considerably greater than the fats recovered as such plus those incidentally retained. The amounts ingested by animals and by human beings would be worth knowing, but they would exclude large quantities that are wasted in various ways on the farms and in households and public eating places.

Furthermore, a distinction may be drawn between intermediate and final consumption. Large amounts of fats are contained in cereals, nuts, milk, oil cake, and other products fed to animals; these are in part used up by the animals, and in part are converted or stored up and appear in dairy products or slaughtered animals. Hence a total of the fats consumed by animals and those derived from animals would contain, in effect, considerable duplication. The difficulties are enhanced by the fact that our statistical information is defective especially in respect to the data on animal fats.

For the United States, the number of animals killed in inspected slaughterhouses is known; the total slaughter at wholesale is reported by the census of manufactures; but only a guess can be made at the number of animals in other uninspected, principally rural, slaughter. Moreover, there is no clear-cut idea of the fat content per average animal slaughtered; the long-time trend seems to be downward, because younger and lighter animals are coming to slaughter in response to the increasing demand for lamb instead of for mutton, for bacon-type hogs instead of for the lard type, and for baby beef instead of for three- to four-year-old steers.

To take even the best ratios derived from packing-house practice and apply them to the total wholesale slaughter of cattle, calves, sheep, and hogs as reported in the biennial censuses, and also to the farm and retail slaughter for which the data are officially admitted to be crude estimates, would be little better than intelligent guessing. At the best, there are gaps in the information so large as to make statistical conclusions on several important components, and on fats and oils as a whole, exceedingly unreliable. It would be of interest to know, for each of the several sources of fats:

- The amount wasted for lack of any attempt at recovery;
- The amount ingested;
- The amount wasted in households and public eating places;
- The amount fed to animals; and
- The amount used in industrial production other than food.

We should like these data for fats produced, fats imported, and fats exported. We should like to distinguish between gross and net consumption. Unfortunately, for many of the sources of fats the statistical information is exceedingly defective and it is difficult to reach reliable bases upon which to make the estimates necessary for arriving at supplementary data. We have made attempts to reach a rough approximation to the amounts of fats and oils available for human consumption or industrial use in the United States in recent years, but at several points gaps are so wide and the procedure is so open to objection that we do not feel justified in presenting here even a preliminary approximation, lest in spite of reservations and qualifications it should be taken for more than it would be worth.

In the discussion that follows, therefore, we have incorporated quantitative data only to a very limited extent. The two principal items omitted, fat of dressed meats and milk fats (including butter), are so difficult to estimate that we have not included them; yet their sum probably exceeds the sum of the fats and oils here itemized. They go largely into edible uses, except as milk is fed to animals, though in considerable measure meat fats are wasted in the household or subsequently recovered from household wastes for industrial use. The principal other items omitted are fats incidentally consumed in fish, poultry, grains, vegetables, and fruits. Judging from estimates of Raymond Pearl for the period, the sum of these items is probably less than a billion pounds a year, exclusive of the fat in grain fed to cattle.

Fats and Oils in Diet

The function of fats and oils in the diet is mainly to furnish energy to operate the animal machine. In the body, fats are burned as truly as though they were burned in a candle or under a steam boiler, and the end-products of the combustion are the same—carbon dioxide and water. Moreover, the amount of energy they furnish is the same, namely from 9 to 9.4 calories to the gram, whether they are burned within or outside an animal body. (A large calorie, which is the one here used, is the quantity of heat necessary to warm 1 kilogram of water from 0 deg Centigrade to 1 deg Centigrade. A small calorie is the quantity of heat necessary to warm 1 gram of water from 0 deg Centigrade to 1 deg Centigrade.) They yield more energy than the other important classes of foodstuffs, such as carbohydrates and protein (albumin) Conditions and Trends of Production—Vegetable oils major products), which furnish from 3.8 to 4.2 calories to the gram. Neither fats nor carbohydrates ingested are wholly burned at once unless they are needed for the operation of the animal machine.

If not so needed, carbohydrates are for the most part completely changed in character by conversion into the fat characteristic of the species. This is stored in the body as a reserve against the possibility of a future period of food shortage. Indeed, most of the fat of domesticated animals is produced from starch. Thus the lard of the hogs of the corn belt is mainly derived from the starch of corn (maize).

It has the characteristics of fat normal to the animal, is normally stiff, and hence is especially prized. Unlike ingested carbohydrate, the fat of the food, if it is not at once burned, is changed comparatively little. For the most part it is deposited in but slightly modified form with other fat, made by the animal from carbohydrate or protein, in the storehouses for fat—the adipose tissue under the skin, in and about the viscera, and elsewhere. Therefore, if the food fat differs in its properties from the fat natural to the animal and if there is a great deal of it in the feed, the storage fat formed by the deposition of food fat gives an abnormal character to the fat which is obtained from the animal after slaughter. The dietary fats may be divided into four groups:

- Those consumed as such on the table;
- Those employed in the preparation of foods;
- Those consumed incidentally in the ingestion of meats and other animal products; and
- Those consumed incidentally in cereals, legumes, fruits, and vegetables.

The fats consumed as such on the table in the United States are principally butter, butter substitutes (chiefly coconut and oleo oil), and salad oils (principally cottonseed, corn, and olive oils). The fats and oils used in the preparation of food are butter, lard, lard compounds (for the most part cottonseed oil), cottonseed oil as such, corn oil, peanut oil, butter substitutes, and, in certain types of confectionery, coconut oil and cacao butter. Fats and oils consumed incidentally in the ingestion of meats and dairy products are found principally in milk, cheese, poultry, eggs, beef, mutton and lamb, pork, and fish.

The fat of meats is of course to a material extent not ingested but wasted except in so far as it is recovered in garbage and similar greases. The fats of the cereals, legumes, fruits, and vegetables as a class are important in the aggregate nutritionally, but have little direct commercial importance, since few foodstuffs are purchased on the basis of fat content. Practically all foodstuffs except sugar, water, and certain condiments contain some fat. Fresh fruits and vegetables contain merely traces; nuts, on the other hand, are rich in fats, as are also chocolate products. Oat meal and corn meal that is made without removing the germ are relatively rich in fat; wheat flour is poorer.

A study of the trends of consumption encounters the difficulty that statistics of fat in the diet over a series of decades are not available. Certain inferences seem nevertheless warranted. Trends of consumption may be grouped in two classes according as they are due primarily to changes in habits or primarily to substitutions made by manufacturers of which the consumer may or may not be aware. The two classes of course overlap, for in certain cases both factors play a part. Thus a manufacturer may create a new product which leads to a new food habit or, vice versa, consumer demand may lead to substitution or creation of a new product by the manufacturer.

Influence of Changes in Food Habits

Change of food habits may be of two sorts. It may be purely quantitative or it may be qualitative. By a quantitative change is meant increase or decrease in consumption of a given fat rather than the substitution in the diet of one fat for another. By qualitative change is meant primarily a substitution of one fat for another. However, since for physiological reasons the intake of food is practically constant for any individual under any given set of circumstances, a decrease in fat ingestion usually involves either a substitution for it of some other kind of food, or of some other kind of fat. Conversely, an increase usually involves a decrease in consumption of some other kind of food or of fat. Since all fats and oils have very nearly the same food value, one can be substituted for another without change in the volume of food ingested. This statement applies to energy values. Certain fats contain small quantities of chemical substances of unknown chemical nature but of great importance to health. These are known as vitamins or food accessories.

Different fats contain different amounts of them depending upon their origin, method of preparation, and other factors. From the point of view of their vitamin content all fats are not of equal nutritive value; but these are considerations. However, if another food be substituted, the volume of the diet is thereby necessarily increased since no other food has, weight for weight, so great an energy value. It follows that if the replacement of fat in the diet by other foods goes too far the diet must become very bulky if it is still to furnish the same energy value, and this fact sets mechanical limits to the substitution of other foodstuffs for fat in the diet.

Since the caloric requirements of an individual, as stated above, are constant under given conditions, it follows that under these circumstances a reduction in consumption of a fat involves the substitution for it of either an equal amount of some other fat or a corresponding amount of some other foodstuffs. But conditions do not remain constant for any person. He grows old and requires less food. He may become stouter or leaner, then consuming either more or less food. He may change his means of livelihood from one requiring hard manual labour involving high food requirements to a sedentary occupation requiring a low food intake. He may migrate from a very cold climate where food requirements are relatively high to a hot one where they are relatively low. In any nation all these changes are taking place in some individuals in one direction, in others in the opposite direction. For short periods, the result is a reasonably constant consumption.

Over a long period of time this is no longer true. The age distribution of populations changes. If the birth-rate falls and the average span of life lengthens, the proportion of old people who consume little food increases, the proportion of young people who consume much decreases. If it becomes unfashionable to be stout, less food is consumed by the nation as a whole. If the proportion of manual labourers becomes less and that of machine tenders and sedentary workers greater, the per capita consumption of food tends to fall. Since fat is the most concentrated form of food energy, its intake tends to be reduced rather more than that of other forms of food.

In the United States exactly these changes have been taking place for decades; and there is evidence of a corresponding decline in the nutritional use of fats and oils, as part of a general reduction in the per capita food requirements. This reduction is the result of substitution of machine labour for man labour, the decline of outdoor employment in severe winter weather, the improved heating of buildings, the trend to lower average body weights of the people generally, and changes in the age distribution of the population. The population is acquiring more sedentary characteristics, needs less food, and the average per capita requirements of foodstuffs in terms of calories are automatically reduced. Most of these factors are the result of the economic evolution of the country and in essence reflect improvement in the standard of living.

The rise of the standard of living is responsible for certain other changes in food habits. It results in a diversification of the diet with increase in the use of dairy products, fruit, vegetables, and sugar, and decline in the ingestion of cereals and fats. This general statement does not hold for the fat of milk, the use of which is on the increase. Fat consumed incidentally to the ingestion of meats and dairy products has always been heavy. The consumption of milk and poultry fat is increasing, that of beef, sheep, and hog fats is declining. Plain cooking is being replaced by more fancy cooking, which includes more discriminating culinary uses of fats and oils. American practice has departed from the British custom of boiling vegetables without fat. Much of the crackers and bakers' bread consumed now contains shortening.

These facts might seem to suggest increased use of fats and oils. A survey of the entire field, however, suggests that the net result is a lower per capita ingestion of fats in general. The heavy fat rations of hard workers are a thing of the past—fat-backs and sow-belly are no longer staples. We use more butter, but less hog fat. The current public taste for younger and lighter-weight animals represents a substantial reduction in fat. It seems fair to infer that, milk fat and poultry fat aside, the decline in the ingestion of fats applies to animal fats. Over a generation, indeed, it is possible that there may have been an absolute increase in the per capita consumption of vegetable oils with an absolute decline in the per capita consumption of animal fats outside of milk fat.

Influence of Changes in Food Manufacture

The changes in trends of food-fat consumption are not all by any means due solely to change in habits. Some of those that are quantitatively extremely important are due to substitutions on the initiative of manufacturers of food-fat preparations—for example, the substitution of lard compounds for lard—or to the increasing availability of new types of fat—for example, coconut oil. Some of the possibilities of substitution in food uses have already been touched upon incidentally.

However, since dietary consumption is the major and the higher or premium use of fats, the practice and possibility of substitution require elabouration because they are basic to any consideration of the commodity economics of this important group of raw materials. The substitution of one fat for another, indeed the mere possibility of such substitution, naturally has the greatest influence upon prices.

Given sufficient price inducement, one fat may in some cases displace another with the widest repercussions upon the producer, the farmer, the trader, and the manufacturer. It is one of the purposes of the *Fat and Oil Studies of the Food Research Institute* to present studies from time to time upon far-reaching movements of this general character. That considerations of price, costs of conversion and refining, lack of technological skill, and legislation in the interests of public health and sanitation limit the volume of inedible fats turned to edible uses, has already been pointed out. But there are other factors as well that limit not merely the diversion of inedible fats to edible uses but also the substitution of one edible fat for another. The most prominent of these are greater or lesser adaptability of different fats to use in the preparation of different foodstuffs, and psychological and sentimental factors that in the food industries play a greater part than in the arts.

Consumption of food products

The consumption of food in India was estimated at approximately INR 4,000 bn at 1993-94 prices (INR 7,250 bn at 2001-02 prices) in the year 2001 -02, growing at a CAGR of 7.8per cent (1996-2002). Food consumption has been driven primarily by increase in per capita expenditure on food items. In the period 1996-2002, the per capita expenditure grew by 6 per cent per annum and population increased at the rate of 1.6per cent per annum. Urban food consumption has grown at 9.1 per cent annually and is estimated at INR 1,540 bn (1993-94 prices). Rural food consumption, on the other hand, has grown at 7per cent annually and is estimated at INR 2,550 bn (1993-94 prices). Uttar Pradesh is the largest consumer market for food products followed by Maharashtra, West Bengal, Bihar and Andhra Pradesh. Among the fastest growing markets are the North East states, Gujarat and Andhra Pradesh.

The share of grain-based products is the highest in the Indian consumer's food basket, followed by milk and milk products, vegetables, edible oil and meat products. The growth rates for fruit, vegetables, meat and dairy products is higher than grains and pulses, indicating a shift in consumption pattern. This also implies the need for diversification in agricultural production to match the change in consumption patterns.

Consumption of processed food products

Table. Estimated Total Food Market in 2003-04 (INR bn) at current prices

Total food Consumption*	8,600
Processed foods**	4,600
Primary processed food (includes packed	2,800
packed milk, unbranded edible oil,	
sugar, pulses, spices and salt)	
Value-added foods (includes processed	1,800
juices, jams, pickles, squashes,	
products - ghee, paneer, cheese, butter,	
branded edible oil; breads, biscuits,	
foods; processed meat, poultry and	
and chocolates; alcoholic beverages - beer,	
and malted beverages)	

Table. Level of Processing in Perishable Products

Products		Level of	
	Organi	Unorgan	Total
Fruits and	1.4%	0.8%	2.2%
Milk & milk	13%	22%	35%
Meat and Poultry			
Buffalo	21%	-	21%
Poultry	6%	-	6%
Marine	8%	-	8%

The market size for processed foods for year 2003-04, at current prices is estimated at INR 4,600 bn, which is about 53 per cent of total food consumption. This includes processing in unorganized sector in dairy (Halwais) and grains (chakkis).The market size excluding these segments is about INR 3,300 bn. The market size of processed foods at factory cost for 2003-04 is estimated at INR 2,100 bn[3] implying about 60 per cent mark-up from factory cost to market price. However, primary processed products constitute as much as 62 per cent of processed foods consumed, with value-added products being the balance 38per cent.

'excluding consumption of alcoholic beverages and out-ofhome consumption

"excluding out-of-home consumption but including alcoholic beverages and processing in unorganized sectors in dairy (halwais) and grain milling (chakkis)

The low share of value added products in food consumption is due to low level of processing across agricultural products.

The key impediments to growth of processed foods include demand-side, supply-side and regulatory issues. The demand side factors include:

- Low per capita income
- Socio-cultural factors such as preference for freshly cooked products as compared to packaged products and availability of low-cost domestic help

The supply side factors constraining the demand for processed foods include:

- Long and fragmented supply chain with independent players engaged in various value-addition activities such as input supplies, cultivation, processing, distribution, storage and retailing (unlike in other major agriculture/food producing countries, where a large number of food companies are integrated across the chain. *e.g.* ADM, Cargill, Tyson Foods)
- High cost of raw material (due to low productivity and poor agronomic practices) and lack of varieties amenable for processing
- Lack of scale

- High cost of packaging
- Lack of infrastructure (cost of power, lack of cold chains, storage, handling and transportation)
- High cost and poor quality of distribution.

CHEMICALS IN FOOD PROCESSING

The same over commitment to potentially risky methods is found in other phases of food production. Complex problems of storage, cleansing, handling, refining, cooking, mixing, heating and packaging are often solved by the use of chemical additives. A half century ago the yellowish tint in freshly milled flour was removed slowly by aging; today it is removed rapidly by oxidizing agents. The miller no longer has to store flour and wait for it to mature. Emulsifiers are employed, partly to "improve" the texture of a product, partly to speed up processing. Their "functional advantages" in ice cream, for example, are that the "mix can be whipped faster, the product from the freezer is dryer and the mix has less tendency to 'churn' in the freezer."

Chemical antioxidants are used to help prevent oils and fats from becoming rancid, a problem that becomes especially pronounced when natural antioxidants are removed during large-scale processing. Chemical preservatives are added to food to permit longer storage in wholesale depots and to extend the "shelf life" of a product in retail outlets. In time chemicals are turned from mere adjuncts of food production into "technological necessities." Their use may result in new machines and facilities, the abandonment of old processing methods and a broad reorientation of technology to meet the new chemical requirements.

The fact that a chemical is technologically useful does not mean that it is biologically desirable. The two phrases are occasionally used interchangeably, as though they were synonymous. Some of the chemicals most widely used in the production of food have been found to be harmful to animals in labouratory tests. Nitrogen trichloride, or agene, was employed for decades as a bleaching and maturing agent in flour before it was discovered, in 1946, that the compound produces "running fits," or hysteria, in dogs.

The use of agene in the bread industry has since been discontinued. Although there is no evidence that agenized flour causes nerve disorders in man, no one can say with certainty that cereal products treated with agene are completely harmless to consumers, especially if eaten day after day for twenty years or more. The toxicant in agenized flour "produces nervous disturbances in monkeys, but not epilepsy in man," observed the late Anton Carlson, one of the most distinguished physiologists of his day. "That is not sufficient for me, because there are many other types of nervous disturbances that may be aggravated or introduced by this chemical."

Another group of chemicals, notably certain types of polyoxyethylene derivatives, were widely used in the United States as emulsifiers before they were banned as potential hazards to consumers. Former F.D.A. Commissioner Charles W. Crawford is reported to have described the agents as "good paint removers." According to data furnished to the Delaney Committee, rats that had been fed two commercial preparations of polyoxyethylene-lb/> derived emulsifiers showed blood in their feces and developed a variety of kidney, abdominal, cardiac and liver abnormalities.

These preparations were used primarily as bread softeners but also found their way into other foods. The polyoxyethylene derivatives constitute a highly suspect family of emulsifying agents, yet some of them are still added to cake mixes, cake icing, frozen desserts, ice cream, candies, dill pickles, fruit and vegetable coatings, vitamin preparations and many food-packaging materials. Similar examples can be culled from other groups of chemical aids. In general, the toxic effects of many commonly used compounds are difficult to determine, but individually and in groups they have aroused suspicion and concern. "The fortification of oils and fats against oxidation and rancidity is another field in which there are serious questions of possible toxicity," warns F. J. Schlink, technical director of Consumers' Research and editor of *Consumer Bulletin*. "The natural antioxidants which delay rancidity are lost during the factory processing of refined oils and fats; then the attempt is made to restore the antioxidation qualities by addition of fat-stabilizing substances. Most of the materials that have antioxidation properties are known to be toxic to some degree."

The term "technological aid" has been used loosely to include food additives that could be dispensed with entirely if a manufacturer improved his processing methods and the quality of his product. The question of whether a chemical, instead of being legitimately employed to furnish distant consumers with a perishable product, was being used rather as a device for masking inferior ingredients and unsanitary methods arose in sharp form early in the century, when sodium benzoate was employed extensively as a preservative in catsup. Harvey Wiley, the first administrator of the federal food and drug law, regarded sodium benzoate as a hazard to consumers.

Supported by data from feeding experiments performed with human beings, Wiley claimed that the preservative causes injury to the digestive system. He was convinced that catsup need not deteriorate readily if clean facilities and proper ingredients are used in preparing it. Challenged on this point by Charles F. Loudon, a canner in Terre Haute, Indiana, Wiley appointed a bacteriologist, Arvill W. Bitting, to work at the Loudon factory and prepare a formula for a stable catsup preparation without artificial preservatives. Bitting was successful. Moreover, after inspecting numerous canning factories and examining many brands of catsup, Bitting concluded that manufacturers who needed sodium benzoate often used inferior products and were careless about sanitation. "The whole spirit and tradition of the pure-food law is against the use of preservatives and substitutes," declared Mrs. Harvey Wiley more than forty years after the Loudon episode. "The intent of the law is to encourage the manufacture of high-grade ingredients, not to try to hide inferiority and cheapness by the use of chemicals and preservatives." If this is so, the spirit of the law has been ignored for decades.

The most up-to-date surveys of chemical additives in foods show that preservatives are added to cheese, margarine, cereal products, jams, jelly and many other processed foods. Uncooked poultry is dipped in solutions of antibiotics. Aside from the damage to public health that may arise from the use of questionable chemicals as preservatives, the longer "shelf life" acquired by some of these foods may cause a loss of valuable nutrients. Many chemical additives in foods perform no technological or nutritional function whatever. They cannot be regarded as materially useful by any stretch of the imagination. These chemicals enhance the colour of a food or make certain products feel soft and newly prepared; in some instances they have been used to replace costly but valuable nutrients with inferior ones. In the least objectionable cases, an additive will deceive the consumer without impairing his health.

The application of an innocuous vegetable dye to some foods often leads the consumer to believe that he is acquiring a better, more wholesome, or tastier product than is actually the case. In the most objectionable cases, a chemical is toxic to a greater or lesser degree. This type of additive not only serves to change the appearance or conceal the nutritive deficiency of a food; it exposes the consumer to a certain amount of damage.

Nitrates and nitrites, for example, are used to impart a pink colour to certain brands of processed meat. Ostensibly, this is their sole function. The addition of these chemicals to meat, especially frankfurters and hamburger meat, would be objectionable if only because of the deception which makes it difficult for the housewife to distinguish between high-quality and low-quality products. But this is not the only deception perpetrated on the consumer. Used together with salt, nitrite compounds definitely extend the "shelf life" of processed meat products.

When nitrites are ingested, they react with hemoglobin in the blood to form methemoglobin and, like carbon monoxide, reduce the hemoglobin's capacity to carry oxygen. An individual who consumes three to four ounces of processed meat containing 200 parts per million of sodium nitrite (a permissible residue) ingests enough of the compound to convert from 1.4 to 5.7 per cent of his hemoglobin to methemoglobin. Ordinarily, this percentage is insignificant. But if the same individual is a heavy smoker and lives in an air-polluted community, the cumulative loss in oxygen-carrying capacity produced by the nitrites in the food and by the carbon monoxide in tobacco smoke and motorcar exhaust can no longer be dismissed as trivial.

Sodium nitrite is highly toxic in relatively small amounts. About four grams of the compound constitutes a lethal dose for adults. Although Lehman regards 200 parts per million of sodium nitrite as a safe residue, he notes that "only a small margin of safety exists between the amount of nitrite that is safe and that which may be dangerous. The margin of safety is even more reduced when the smaller blood volume and the corresponding smaller quantity of hemoglobin in children is taken into account. This has been emphasized in the recent cases of nitrite poisoning in children who consumed weiners and bologna containing nitrite greatly in excess of the 200 ppm permitted by Federal regulations. The application of nitrite to other foods is not to be encouraged."

Emulsifiers have been used in bread not only as softening agents, which can give stale bread the texture of newly baked bread, but as substitutes for nourishing ingredients. "The record of the bread-standards hearings contain evidence of distribution among bakers of advertising material advocating the replacement of fats, oils, eggs and milk by emulsifiers," George L. Prichard, of the Department of Agriculture, told the Delaney Committee. "The use of such products as components of food may work to the disadvantage of our farm economy by displacing farm products normally used. The record indicates that natural food constituents, such as fats and oils, probably will be reduced in many commercial bakery products if bakers are allowed to employ these emulsifiers."

This substitution affects more than 117 the 117 farm 117 economy, however; it also works to the disadvantage of the consumer. To illustrate this, the Delaney Committee report compared the ingredients in two cake batters prepared by the same company eleven years apart; during the interval emulsifiers were added to the product. The first batter, made in 1939, did not contain synthetic emulsifiers; the second, prepared in 1949, did. "On a percentage basis, the cake batter in 1939 contained 13 per cent eggs and 8.6 per cent shortening.

In 1949, the cake batter contained 6.3 per cent eggs and 4.8 per cent shortening, with somewhat less than 0.3 per cent of synthetic emulsifier."

The report noted that a "synthetic yellow dye could be added to provide the colour formerly obtainable through the use of eggs. The utilization of synthetic yellow dye in commercial cake was practiced before the war. There are indications that the use of artificial colouring matter is increased when quantities of whole eggs or egg yolks are reduced in commercial cake formulas." To heighten the insult, among the most commonly used emulsifiers in 1949 were the polyoxyethylene monostearates. The yellow dye referred to in the Delaney Committee report was probably FDC Yellow No. 3(Yellow AB), which, until fairly recently, was added to many yellow cake mixes. The dye often contained impurities of a potent carcinogen and its use in foods was forbidden in 1959. For a number of years, however, both the emulsifier and the dye undoubtedly appeared in many brands of cake and reached large numbers of unsuspecting consumers.

Problems of this kind are not likely to disappear unless there is a basic change in the viewpoint of the F.D.A. "Inherited from the Wiley era is a too common misconception that all 'chemicals' are harmful and the related idea that any amount of a 'poison' is harmful," the F.D.A. observes in a brochure on food additives. "The fact is, of course, that chemical additives, or food *additives* as they are now being called, have brought about great improvements in the American food supply. Additives like potassium iodide in salt and vitamins in enriched food products are making an important contribution to the health of our people and yet it is a fact that both iodine and some of the vitamins would be harmful if consumed in excessive amounts. Many similar examples could be given to refute these common misconceptions."

This argument is grossly misleading. Iodine and vitamins are indispensable to human life, but coal-tar dyes and benzoic acid, for example, are not. If we adhered to a well-balanced diet of properly prepared natural foods, iodine and vitamins would never enter the body in toxic amounts. Coal-tar dyes, on the other hand, are suspected of being harmful to man in nearly any amount if consumed repeatedly and the kindest statement that can be made for the presence of benzoic acid in food is that the compound is "safe under the conditions of its intended use."

The formula "safe under the conditions of its intended use" exposes the consumer to serious risks when it opens the door to the use of food additives whose biochemical activity is not understood. Many unexpected problems may arise when such additives appear in food. A particular additive may seem to be relatively harmless to the organs of the body, but it may be carcinogenic on the cellular level of life. Another additive may produce insignificant or controllable effects when studied in isolation; brought into combination with various chemicals in food, however, it may give rise to toxic compounds. The body, in turn, may make a toxic additive more poisonous in the course of changing it during metabolism. "At one time it was generally believed that whenever a toxic substance was absorbed the body was capable of calling upon special mechanisms for detoxifying the toxicant," Lehman observes.

"It was believed also that the metabolic pathway that the toxicant followed always proceeded in the direction of the formation of a less toxic compound. Later work on the metabolism of drugs and toxic substances showed that special mechanisms did not exist, but that the toxicant was subject to the same metabolic influences as those which normally operate in the body. The assumption that the metabolic product was less poisonous than the parent substance from which it was derived is also unwarranted simply because in many instances the toxicity of either the original substance and its conversion product is unknown. In other instances the metabolic product is even more poisonous than its parent. The conversion product, heptachlorepoxide, which is two or three times more poisonous than the parent substance, heptachlor [a widely used insecticide], may be cited as an example of this."

Finally, food additives may cause allergic reactions that are likely to go unnoticed for many years. The reactions need not be severe to be harmful. Otto Saphir and his colleagues at the Michael Reese Hospital in Chicago have recently suggested that the development of arteriosclerosis may be promoted by the sensitivity of the body to allergenic compounds, notably certain antibiotics. By using sulfa drugs to produce allergies in forty-two rabbits, the researchers were able in eight months to cause degenerative changes in the arteries of thirty-one of the animals.

These changes closely resembled arteriosclerosis in man. According to a press account of Saphir's report, the rabbits' reactions "were not apparent on the surface." It would be very imprudent to assume that such effects are produced only by chemicals that cause severe or noticeable allergies. If the data of Saphir and his colleagues are applicable to man, additives with even minor allergenic properties cannot be dismissed as harmless.

Today more than 3,000 chemicals are used in the production and distribution of commercially prepared food. At least 1,288 are purposely added as preservatives, buffers, emulsifiers, neutralizing agents, sequestrants, stabilizers, anticaking agents, flavouring agents and colouring agents, while from 25 to 30 consist of nutritional supplements, such as potassium iodide and vitamins. The remaining chemicals are "indirect additives"—substances, such as detergents and germicides, that get into food by way of packaging materials and processing equipment. Many chemical additives are natural ingredients, but a large number are not.

The artificial substances that appear in food range from simple inorganic chemicals to exotic compounds whose biochemical activity is still largely unknown.

Distribution of foods in India

Food products are sold through over 5 million food and grocery stores in India. The organised food retail market is estimated at INR 20 bn (approximately 0.2 per cent of food expenditure in India).This is in contrast to developed countries, where food distribution is highly consolidated. The majority of food and food products are retailed through neighbourhood kirana stores. The kirana stores focus on dry food products in the absence of infrastructure for cold storage. Bulk of fresh produce is sold by vendors with push carts. Meat, poultry and marine products are primarily sold by small retailers in wet markets. Such produce is associated with low product quality, lack of variety and low hygiene levels.

Internationally, food retailers have played an important role in improving supply chain efficiencies such as developing storage and transportation infrastructure, training supply chain members on food hygiene and standards and providing scientific know how to farmers. The key impediments to growth of organised food retailing in India include lack of infrastructure, technology and capital.

Foreign Direct Investment (FDI) is not allowed in retailing with the exception of cash and carry formats. This restriction is based on the premise that entry of large international players will displace existing employment and reduce bargaining power of farmers. However, there is empirical evidence from other countries (USA, China) which highlights that opening of the retail sector to FDI has enabled employment creation, disintermediation, increased farmer access to information and wider choice and cost savings for the consumer.

Exports of agricultural and food products

India has 1.5 per cent (INR 360 bn in 2003-04) share of global agricultural exports (approx USD 522 bn), despite its production leadership in agriculture. India's exports primarily constitutes commodity and primary processed items, where price realisations are low. In addition, many products are showing single digit or negative growth. The reasons for India's insignificant share in global trade include supply side factors such as lack of consistency in supply and quality, lack of cost competitiveness, and demand side factors such as non-tariff barriers, short product life cycles and perception of Indian food products in overseas markets. Most exporters from India lack scale- for example the largest fresh produce exporter records annual sales of about INR 500mn.This has resulted in lack of economies in operations and renders them uncompetitive. Hence, exporters are not able to establish themselves as long-term players in the export market, rely on opportunistic businesses and are consequently, unable to develop technical and managerial expertise. These factors cumulatively lead to low investments by exporters in brand building, quality improvement and brand development.

Status of food distribution in India

Food accounts for the largest share of consumer spending. Food and food products account for about 53per cent of the value of final private consumption estimated at INR 8600 bn (2003-04 at current prices). Food products are sold through over 5 million food and grocery stores in India. The size of the organised food retail segment in India is estimated at INR 20 bn (approximately 0.2 per cent of food expenditure).This is in contrast to most of the developed countries, where food distribution is highly consolidated.

In countries such as Australia, France, Germany, Denmark and The Netherlands, the share of top the 10 retailers is more than 80per cent (USA ~ 35per cent, China ~ 15per cent).

Most organised retailers in India are regional and use single formats such as convenience stores, supermarkets, hypermarkets or cash and carry. In contrast, most international retailers have multiple format models. Besides reflecting the stage of evolution of retailing in India, this also highlights the capital constraints of existing players in India, restricting their ability to make large investments.

In India, food companies are much larger in size than the organised food retailers in India. In contrast, the retailers are larger in scale, than food companies, in developed countries. Given the channel power of retailers in these markets, food companies have started establishing partnerships with them to develop and test new products, share consumer information and undertake consumer promotions.

Key Drivers

Increasing need for convenience: The Indian consumer visits about eight to ten outlets to purchase various food products which constitute the daily consumption basket. These outlets include neighbourhood kirana stores, bakeries, fruit and vegetable outlets, dairy booths and chakkies (small flour mills). With changing lifestyles, there is growing paucity of time, and convenience in food shopping is emerging as an important driver of growth of one-stop retail formats. Availability of quality retail space: Until the late 1990s, the high cost of real estate meant that organised food retail business models were not financially viable in metropolitan areas.

In the last few years, various factors have led to increased availability of real estate for organised retail formats. About 300 malls are at various stages of construction, across metros and mini-metros in the country. The average size of a mall is about 100,000 sq.ft. About 25 malls each are planned in Delhi and Mumbai, primarily in suburban and satellite areas. This will translate into additional retail space of 30 to 40 million sq.ft over the next three to five years.

Key Impediments

The key impediments to organised food retailing include lack of infrastructure, technology and capital. Lack of infrastructure and technology: Food distribution in India is characterised by a high degree of complexity given dispersed production on the one hand, and variance in demand across locations on the other. There is a compelling requirement for appropriate infrastructure for storage and transportation such as temperature controlled warehouses and vans. However, the limited scale of operations of retailers has restricted their investment capacity in these areas.

The recent entrants who have fuelled growth of organised formats are attempting to address this issue, but given the large requirements of capital, the process of transformation is gradual. Further, entry of organised retailers, who have larger investment capacity, will also address the issue of creation of infrastructure for storage and transportation.

Lack of capital: Organised food retailing is a capital intensive business with a long gestation period (typically 5-7 years). Investment is required in buying/leasing land, furniture and fixtures, IT systems, quality control systems, vendor development and infrastructure for warehousing, storage and transportation.

The development of organised food retailing in India has been constrained due to lack of capital. Foreign Direct Investment (FDI) is not allowed in retailing with the exception of cash and carry formats.

Impact of organised Retailing on Employment

The restriction of FDI in the retail sector is based on the premise that entry of large international players will displace existing kirana stores and impact employment. However, there is empirical evidence to suggest that development of large retail formats will translate into greater employment as witnessed in other countries such as China, Japan, Singapore and Malaysia. The sector is labour intensive and contributes significantly to employment. The current share of organised retailing and share of retailing in total employment is tabulated below. Higher share of organised retailing has led to higher share of retailing in total employment. In India, the share of employment of organized food retailers is insignificant, on account of the limited size of this segment at present. Traditional kirana outlets employ approximately 20 million people. However, there is significant shadow employment at these outlets, reflecting that the real employment potential of these outlets is far lower.

India's position in world trade.

Global agricultural exports (including food products) increased from USD 326 billion in 1990 to USD 520 billion in 2003. Strong expansion in food and agri exports in the mid 1990s, was followed by a decline from 1997 to 2000.

This was mainly due to a slump in prices for various agricultural commodities, on the back of demand-supply mismatches. Since 2001,food and agri trade has been growing. The main items in food and agri trade include fruits and vegetables (-USD 75 billion),cereals and cereal-based preparations (-USD 58 billion), marine products (-USD 57 billion), meat and meat-based preparations (-USD 46 billion), milk and milk products (-USD 33 bn) and beverages (-USD 40 billion).Latin America has by far the largest agricultural trade surplus, followed by Oceania. Brazil and Argentina contribute significantly to the Latin American trade surplus. Asia has the largest trade deficit of all regions, primarily on account of Japan. Western and Eastern Europe as well as Africa have a trade deficit too. About half of global trade flows are regionally contained, partly as a result of regional agreements and partly due to the perishable nature of agri-commodities.

There has been a sharp increase in regional trade agreements (RTAs) over the past decade. A total of 259 RTAs have been notified to the WTO by the end of 2002, although only 176 RTAs are operational. An additional 70 RTAs are estimated to be operational, but not yet notified by WTO and about 70 are under negotiation. A well known example of regional trade is in the European Union, where about 75per cent of agricultural trade is within the region.

The intra-regional trade figures for other regions are 33per cent (NAFTA),63per cent (Asia) and 15per cent (Latin America and Africa). The group of G-20 countries, a grouping of developing countries led by Brazil, China, India and South Africa have sought further policy reforms and market liberalization on the part of the developed world. At the Geneva meeting of the WTO, the members have agreed on a further reduction of overall trade distorting domestic support, the elimination of all forms of export subsidies and a reduction of import tariffs. The EU already anticipated this by drastically reforming the Common Agricultural Policy

(CAP). Besides agricultural policies and trade arrangements, demand-supply trends and the macroeconomic environment also significantly influence trade. Another important factor is the change in currency rates.

Exports

India's exports of agricultural and food products are about INR 360 billion (approx. USD 8 billion) which constitutes about 1.6per cent of total global trade (USD 520 billion). Exports of agriculture and food products have grown at 15per cent annually, vis a vis total exports which have grown at 19per cent, in the last decade. The share of agricultural exports as a percentage of India's total exports has decreased from 19 per cent to 13 per cent in this period. Excluding non-food agricultural products (such as paper, cotton and jute), tobacco and seeds; marine exports constitute the single largest product category in agricultural and food exports from India followed by, rice oilmeals, wheat, cashew, tea and coffee. Most of the agricultural products are primary processed. Many products have shown negative or single digit growth. The top 15 export categories in food and agricultural (except for non-food items and tobacco) from India are listed below. With the exception of buffalo meat, rice and cashew, India's share in global agricultural and food trade is insignificant. Marine products which have the highest share in Indian exports, have a 2per cent share of global trade in this category. Among the key food exports, only non-basmati rice, wheat and buffalo meat has shown double-digit growth. Marine products, basmati rice and processed fruit juices/vegetables have shown single digit growth. Products like cashew, tea, spices, coffee have shown negative growth.

Imports

Against exports of approximately USD 8 bn India's import of agricultural products are about USD 5 bn which translates into trade surplus of approximately USD 3 bn (Year 2003-04). India has abolished restrictions on imports of many agricultural and food products post liberalisation. However, tariff quotas are maintained on some edible oils, maize and milk powder.

There are restrictions on imports of certain fats, oils of animal origin, and beef, based on GATT Article XX that permits countries to restrict imports on religious grounds. Some products, such as wheat, rye, oats, maize, rice, canary seed and other cereals, continue to be traded by the Food Corporation of India. Despite liberalisation of imports, India's imports are growing at about 1 per cent per annum for the last five years. While exports cover a wide range of products, imports are skewed towards pulses and oilseeds. Pulses and oilseeds are the key food items in India's imports (73per cent share) of food items displaying year on year growth. Increasing production of edible oils in India can reduce India's dependence on imports to a large extent. Another category which has witnessed growth in imports is Scotch whisky. The volume growth is about 13per cent and value growth is about 6per cent in the last decade. However, the growth is driven by imports of bulk Scotch, which is bottled in India and/or blended with Indian liquor by domestic manufacturers. Bottled Scotch imports have not witnessed any growth due to the current level of import duties.

Issues hindering exports

India continues to be either absent or at best a marginal player in most of the leading markets for its exports. Indian players have not succeeded in establishing direct linkages with buyers/consumers in importing countries, as a result of which a large proportion of exports are being further processed and re-exported by other countries.

Product

Marine products - Decline in raw material availability:

- Fragmented base of suppliers
- Lack of availability of technology *e.g.* for tuna fishing
- High duties on imports of additives/flavourings for making value added products
- Anti-dumping issues (such as excess use of antibiotics) in key markets

Rice

- Outdated milling technology (using rubber roll sheller) resulting in high share of broken rice
- Poor quality of seeds/non-availability of certified authentic seeds leading to lack of consistency in grain quality
- Low cost competitiveness due to state taxes and MSP regime (for non-basmati]

Cashew

- Lack of raw material availability - high dependence on imports of raw nuts, Interstate barriers
- On movement of raw nuts
- Purchase tax for exports in some states (*e.g.*4 per cent in Tamil Nadu]
- Lack of mechanization of processing leading to higher production costs
- Lack of established quality parameters for raw cashew

Key Issues Hindering Exports

Tea- Export market viewed as a 'residual market' to sell surplus production.

- High cost of production
- Production focused on CTC rather than orthodox, whereas demand for orthodox tea is growing faster than for CTC

Coffee- Low realizations due to exports restricted to green coffee and no exports in roasted/instant/branded coffee. Export cess of INR 500/MT leading to low profitability for exporters

Buffalo meat

Mango

- Absence of policy on permitting rearing for slaughter which translates into lack of traceability a key requirement in several importing countries
- Inappropriate facilities at existing municipal slaughter houses.
- License for setting up private slaughter house is difficult to obtain.
- Prevalence of various diseases particularly Foot and Mouth Disease (FMD)
- Lack of exportable varieties (high fibre content; inappropriate appearance and texture and large size of stone)
- Lack of post-harvest treatment facilities such as for vapor treatment
- Lack of packhouses from farm to port

Grapes

High cost of setting up vineyards:

- High cost for obtaining certification for exports. For *e.g.*, the cost of EurepGap certificate is
- INR. 75000/farmer (including the cost of construction of separate storage space for fertilizers and pesticides etc)

Dairy Products

Lack of quality monitoring mechanism in the supply chain as required by many importing countrie.

- Opportunistic production of SMP with low exportable surplus
- Few local manufacturing facilities of value added products
- Lack of appropriate packaging technology

Guar Gum

Variation in production from year to year (high share of rain-fed areas). Increasing domestic demand from non-food applications such as textiles, paper, pharmaceuticals, oil well drilling.

Uneconomic scale of operations

- Lack of consistency in supply and quality
- Lack of cost competitiveness due to statutory charges, intermediation and wastages/losses Inadequate and inappropriate storage and distribution infrastructure

Non tariff barriers Short product life cycle Lack of brand image

Most exporters from India lack scale- for example the largest fresh produce exporter records annual sales of about INR500mn.The low volume translates into lack of economies in operations and makes exports uncompetitive. Hence, exporters are not able to establish themselves as long-term players in the export market, and rely heavily on opportunistic businesses. These factors cumulatively translate into low investments in upgrading skill sets, product innovation, quality improvement and brand building.

- Low economies of scale high cost and low volume.
- Low investment capacity in product innovation and Brand/Market development.
- Low competitiveness in global markets unable to establish as a long-term player.
- Low efficiencies technological and managerial.
- Countries across the world have successfully addressed these issues to become globally competitive.

Rice, Cashew and Coffee exports from Vietnam

Rice Increased competitiveness through higher productivity: High competitiveness in rice has been achieved on the back of consistent increase in area and productivity for paddy. Cropping patterns have been adjusted to increase land area for planting winter-spring paddy and summer-autumn paddy (from 2.1 and 1.2 million ha in 1990 to 2.89 and 2.35 million ha in 2000 respectively) and reducing acreage under low - yielding winter paddy (from 2.74 to 2.4 million ha). Intensive farming and more advanced technologies have led to a consistent increase in paddy yields from 3.69

tons/ha in 1995 to 4.6 tons/ha in 2003; higher than other Asian countries including India (2 tons/ha),Thailand (2.2 tons/ha), Myanmar (3.2 tons/ha) and the Philippines (2.89 tons/ha).

These measures have facilitated the increase in output from 19.2 million tons in 1990 to 31.3 million tons in 1999, and further to 34.4 million tons in 2003. Cashew - Mechanization of processing technology: Vietnam is the largest cashew producer in South East Asia and the third largest cashew exporter in the world after India and Brazil. Over 90per cent of production is exported. The number of cashew exporters have increased from 16 in 1997, to over 50 in 2003.

The cashew processing industry in Vietnam has made significant contributions to enhance exports. The developments in processing technology have enabled Vietnam to export cashew in processed form. Vietnam has developed "Cover split technology-designed by Vietnamese technicians. This technology is cheap and is able to generate a higher ratio of whole seed. Due to easy availability of efficient technology, the number of processing companies increased from 6 in 1986 to 30 in 1994 (with total capacity of 75000 tons/year) and to 62 in 1999 (with total capacity of 250000 tons/year) to about 120 in 2003. Coffee - Government Support: Vietnam is ranked as the second largest coffee exporter in the world since 2000, next to Brazil. It exports coffee to about 62 countries. The two largest markets include the European Union (47 per cent share) and the USA (15 per cent share).The Government has played an important role in increasing coffee exports. Some of the initiatives include:

Inviting foreign investment in production and trading of coffee - many of the world's largest trading houses in the world such as ED and F Man, Newman Groupe and O Lam are present in Vietnam. Promoted application of international standards to domestic coffee. In order to cope with the market changes, the Government and industry bodies carried out studies to determine the appropriate proportion of Robusta and Arabica to be cultivated in line with global demand. Accordingly, Robusta plantations were replaced by Arabica.The aim is to achieve a proportion of Arabica and Robusta as 75:25,to align it with global demand.

Employment generation potential of Food Processing Industries

Food Processing has significant potential for employment generation not only directly but across the supply chain in production of raw materials, storage of produce and finished products and distribution of food products. For *e.g.* a grant of INR 66.7 million (total investment of approximately INR 250 to 300 million) to 35 units in UP in 2003-04 has resulted in direct employment of 2,500 and indirect employment of 20,000, with a significant rural component. Employment intensity is significantly higher in the Small Scale Industries (SSI) sector as compared to the organized sector for the same level of investment.

The incremental employment in the organized sector in FPI sector by 2015 on the basis of the stated vision is estimated at 8.2 Million. The sectoral break-down is as follows.

The employment intensity in the organised sector is 1.8 direct and 6.4 indirect per million of investment. The ratio of indirect to direct employment is therefore 3.5. Rabo India has estimated investment required in the organised sector of FPI as INR 997 bn in next ten years. Hence the employment generation potential in the orgnaised sector is 8.2 million including 1.8 million direct and 6.4 million indirect for an investment of INR 997 bn. The unorganized sector in food processing requires an investment of about INR 100 bn in next ten years (estimated on the basis of output ratio as 2:1 and capital intensity ratio as 5:1 of organized and unorganized sector. The employment intensity is estimated to be approximately 10 direct employment per INR million of investment in the unorganised sector (Source: Dr.JS Bedi Analysis).

This will lead to direct employment creation of 1 million in unorganized sector. The indirect employment generation in the unorganised sector will be about 1 million (assuming ratio of direct to indirect employment as 1:1). The above analysis assumes no replacement of existing employment.

SSI in Food Processing

The SSI Sector accounts for 95per cent of industrial units in the country,40per cent of value added in the manufacturing sector, 34per cent of national exports and 7per cent of Gross Domestic Product (GDP).The SSI sector is the largest employment generator next only to Agriculture.

It has been estimated that an investment of INR 1 million in fixed assets in the small-scale sector generates employment for forty persons and produces more than four million rupees worth of goods or services.

The Food sector is a leading employer within SSI, providing employment to 480,000 persons (13per cent of SSI).

The SSI sector is less capital intensive with a high potential to generate employment.

However, the efficiency of SSI units is impacted by the following:

- Lack of capital/credit
- Inadequate Training technical/managerial
- Tools and technology (traditional and less efficient)
- Limited market knowledge (demand, food standards)

- SSI's role in semi-processing: The organized large-scale sector is focused on processed foods, where SSI cannot compete due to lack of marketing and distribution strengths. However, SSIs can play an important role in procuring from farmers and primary processing of produce to increase shelf life and make it available to processor/marketers who have access to the final consumer.
- Need for product innovation and branding: There is a strong need to provide necessary training and RandD support to SSIs to promote product innovation. Also, SSIs have limitations in terms of investments on brand development. There is a need to promote public private participation in supporting collective investment by SSIs in branding.

Financing needs of the agricultural and food supply chain

The Indian agribusiness supply chain is highly fragmented, with independent players engaged in various value-addition activities such as input supplies, cultivation, processing, distribution, storage and retailing.

The scale of operations of each entity is limited not only due to the structure of the supply chain, but also on account of Government policies as well as due to demand-related factors.

In contrast, internationally, a large number of food companies are integrated across the chain. *e.g.* players such as ADM and Cargill have business interests in agricultural inputs, procurement, transportation and storage of produce, processing and value addition.

The structural complexion of the Indian supply chain thus translates into limited scale of financing as well as higher risk, given the lack of control of each of the players, on the supply chain.

Issues with farmer financing

It is important to understand the issues faced in farmer level financing, both from the borrower's as well as the lender's perspective, and develop solutions which address these issues comprehensively.

Farmer's Perspective - Limited Access to Credit

Non-availability of timely credit from organized sources, mainly on account of the processes and procedures of banks/financial institutions. This compels farmers to rely on intermediaries.

- Inadequacy in credit availability-The 'scale of finance' method, stipulated by RBI, caps total credit availability per farmer.

- Inability to offer tangible security/collateral cover to access credit from banks Subsistence farmers, who constitute a large proportion of India's farming community, are unable to benefit from the organized credit system since they cannot offer adequate tangible security. Further, the security offered (land documents) are notional and cannot be enforced by banks.
- The above factors, thus translate into continued reliance on intermediaries for financing. While intermediary financing is timelyas they do not require the level of documentation as banks, the cost of funds is exorbitant, nearly 4 times that of interest rates charged by banks.
- A more efficient agri-supply chain requires that farmers' financing needs are addressed by banks and linkages of farmers with middlemen are scrapped.

Banker's Perspective

- Lack of credit discipline among borrowers Lack of credit discipline, to some extent is infused by the Government's decision to waive outstanding loans of farmers. Policy-related uncertainties translate into a huge risk for banks. Instead of loan waivers, the Government should seek to impose a short-term moratorium on repayments, to enable farmers to overcome the adverse impact of a poor monsoon/crop failure.
- Lack of tangible security from farmers- as mentioned above, farmers are unable to provide tangible security to financiers.
- Lack of linkages of farmers with processors- In absence of forward linkages with markets, banks run a huge performance risk, as there is greater potential for farmer default.
- Inadequate insurance cover the existing crop insurance schemes are inadequate, and this further increases the risk profile of farmers
- Norms for priority sector lending- As per the current RBI policy on priority sector lending, kharif credit is not accounted for in priority sector lending. This results in a lop-sided focus on extending financing to farmers, favouring crops cultivated in the rabi season.

Policy-related Issues

- Cooperatives Act: There are about 95,000 Primary Agriculture Credit Co-operatives (PACS) and about 100,000 marketing co-operatives. As per the Cooperative Act, cooperatives are allowed to borrow from Cooperative Banks, DCCBs and RRBs. This provision needs to be extended to scheduled banks, in order to enhance flow of credit to farmers.

- State Warehousing Corporations Act: Private sector scheduled banks are constrained from lending to State Warehousing Corporations. Given the sizeable investment required to create grain storage infrastructure, it would be in the interest of the CWC/SWCs that a wider participation of lenders is permitted.
- Restrictions on land holdings: The State Land Holding Laws prohibit consolidation of land holdings.

Financing to farmers with larger land holdings, and thus larger scale of operations, could result in higher credit flows from banks. Further, the Government could also consider consolidation of wastelands and leasing of these to corporates. This would not only enable increase in financing to agriculture, but also aid crop diversification.

Challenge Facing the food industry

The challenge facing the food industry, and this is linked to shifts in the food system more generally, is how to make healthy foods more marketable, and marketable foods appear more healthy. Both questions are capable of eliciting a response from food manufacturers, but to understand the background to this response we need to examine the patterns of resistance which have grown up to recent changes in the food industry, patterns which echo some of the forms of resistance to environmental. The beginnings of what Belasco calls a 'countercuisine' are linked to a number of quite important social changes which, as we have already noted, shift the individual's attention from the sphere of production relations to those of consumption.

Three aspects require our attention: the way in which we arrive at our impressions of what foods to avoid and what foods to consume (the quality of food as a commodity in the market); the way in which food 'signifies' and expresses other aspects of our lives (food as self-enhancement); and finally, the way in which our consumption of food is linked to broader questions of ownership and organization in the economy as a whole (food as part of a broader political economy). Almost unnoticed, issues surrounding each of these aspects of the consumption of food have assumed enormous importance in recent years, but close examination reveals that what at first appear to be almost random evidence of social resistance to the modern food system have their roots in fundamental social changes.

These social changes need to be outlined, and their implications examined. It is our contention that too much attention has been devoted to the 'signifying' aspects of food and diet, the semiology of food consumption, by sociologists and anthropologists, to the detriment of a wider understanding of the social transformations implied by the economic and technological changes in the food system.

As we have seen, women's roles have been transformed in a number of ways, some of which are clearly irreversible. Domestic work, including housework, has been opened up to important market forces, and the manufacture and purchase of 'white goods' for the home is now an important activity. At the same time the idea that housework is no longer onerous has served the interests of manufacturers and others. It is far from clear that technology in the home is labour-saving; this assumption, needs to be examined critically. Evidence exists that a human price is exacted for the benefits of speed and convenience which 'white goods' bring.

Housework standards change, and women expect (and are expected) to raise the standards of housework in turn. At the same time, it is likely that women's release from the worst forms of labour drudgery is associated with more time being spent by them on other activities in the home, particularly attention to children. This is not to say that many household tasks are not easier; they are. It is merely to point out that the definition of housework, and with it women's responsibilities in the home, is constantly shifting. It is not immutable. It is also clear that since the 1970s important changes have occurred in popular understanding of the relationship between food and health. Preventative medicine may only be in its infancy, and receive little official encouragement, but for some groups of people in the industrialized world, healthy eating is now considered essential.

Healthy eating in the past depended critically on local custom and diet and, most importantly, income. Today this is still true, but local variations have less importance, and additional factors have made their appearance, which have engaged the attention of a battery of professionals, such as health educators, doctors, consumer groups and alternative therapists.

The perceived need to diet to reduce obesity has given rise to a huge consumer market in low-calorie foods, including the controversial 'low-cal' liquid diets and innumerable diet 'systems'. The publication of evidence about heart disease and cancer has led to some marked shifts in the public consumption of certain foods, especially in the United States, where the changes date back to the publication of the McGovern Report in 1977. Food has become one of the obsessions of our time, associated as it is with good health, longevity and the occurrence of stress. Physical fitness has also become an obsession in some quarters, perhaps also a 'fad'. Evidence from the United States suggests that, while in 1960 only a quarter of adults exercised regularly, by 1980 over half did so.

The point is not that everybody is exercising more, but that more people are, and that health is increasingly linked to diet in many people's minds. This is not a trend likely to be discouraged by the food industry, since it also represents a change with commercial possibilities. Indeed, the giant food firms have jumped on this bandwagon, adapting to each new nutritional recommendation—low sodium, low saturated fats, high fibre—to the point of confusing consumers grappling with labels and elusive media-speak, such as 'lite' and 'natural'.

Changes have also occurred in public perceptions of the two processes: appropriation and substitution. Following the scare over DDT between 1969 and 1972 in the United States, and subsequent publicity about the use of other pesticides, attention has been given to the effects of pesticide residues in food, especially by the London Food Commission in the United Kingdom. Few studies have been undertaken, and even fewer given publicity, about the relative nutritional quality of organically produced and chemically produced food. Pesticide use in the United States rose 500 per cent between 1950 and 1986, but in the latter year one-fifth of United States's crops were lost to pests, the *same percentage as in* 1950! The tactic employed by the food industry in meeting the criticism of environmentalists and whole food enthusiasts, has been that the industry can 'compensate' through industrial processes for the shortcomings of nature.

The industry argued that organic foods were more expensive (ignoring the hidden subsidies provided to chemically produced food), that organic foods could not meet market demand, that food additives increased the palatability of food, and that convenience removed the drudgery from food preparation. By the late 1960s in North America, and perhaps a decade later in the United Kingdom, food manufacturers began to 'reposition' themselves, as well as their food products. Now they were on the side of emancipated women and environmentalists; they offered themselves as accomplices in women's drive for more independence and in the vanguard of responsible, sustainable resource management.

Market researchers have credited much of the improvement in the food industry to their own responsiveness to consumer pressure. In the words of an industry spokesperson: For example, ten years ago the only place people could find additive-free *natural* products were in the comparatively cramped and small premises of health food stores. Today such products, not only much improved and—dare we say it?—engineered to satisfy consumer preferences, are mainstream products in prime placing in major supermarket chains the length and breadth of Britain.

This represents one side of the story, but in practice healthier eating has also presented a challenge to the food industry which it has found difficult to grasp. The problem, in a nutshell, is how to leave out food additives and processing, and still add value to the final product.

The post-war period had seen additives 'substitute' for natural ingredients, while much that was natural had been removed from food. Food was given more value added through packaging, processing and ensuring it stayed 'fresh' longer.

The challenge represented by making 'healthy' foods from 'health' foods was how to make profits by appearing to do less to the product. The industry responded to the challenge, as we have noted, by both market and product differentiation—identifying, and exploiting market opportunities for 'niche' products, such as low-calorie foods, ethnic health foods and fitness-related foods.

Finally, the food industry has been in the forefront of the promotion of food supplements; people can now supplement their inadequate diet from a range of commercially promoted food 'accessories'.

In 1985 over US$3 billion was spent in the United States on vitamin and mineral supplements and over $4.5 billion in supplements to fortified breakfast cereals. This diet supplementation was the inevitable outcome of the way eating had changed in the United States. In 1909, 40 per cent of calories had been provided from fruit, vegetables and grains. By 1976 only 20 per cent of calories came from these sources, the rest from fats and refined sugars. As people ate more calories they ate fewer vitamins and minerals—supplements have helped to fill the vacuum left by changes in the 'affluent diet' of North America.

Another important social change that has accompanied the shifts in diet referred to above is that governments in the industrialized countries have been forced to make some accommodation to the pressures put upon them by food consumers, and interest groups representing consumers. In the United Kingdom this has come about largely as a result of successive food 'scares', such as those over salmonella, listeria and bovine spongiform encephalopathy ('mad cow disease') in the last few years.

The extent to which governments will respond to consumer anxieties is still unclear, but the current debate about public nutrition policy and food legislation represents an excellent illustration of the kind of contradiction to emerge from the development of the modern food system.

Quality Control in Food Processing

The maintenance of a well functioning Quality Assurance (QA) programme is essential if a consistent product is to result which meets all required standards. Such a programme should be based on Hazard Analysis and Quality Analysis Critical Control Point (HACCP and

QACCP) systems. HACCP and QACCP are more proactive than traditional approaches to QA/QC activities. The establishment of such programmes is the responsibility of QA personnel but the execution of it involves everyone in the company. To avoid ambiguity regarding responsibility for any QA function, it is important to assign specific HACCP/QACCP accountabilities to responsible persons and groups. A QA programme must consider all activities impacting upon product quality, from raw materials and ingredients used to product handling through distribution channels all the way to the final consumer. In respect of this, Wilson has outlined the following required components of a QA system:

- Raw material control – standard specifications must be adopted for all ingredients which must then be inspected to ensure conformity;
- Process control – all chemical, physical and microbiological hazards as well as quality factors must be identified, critical control points (CCP) must be established, monitored and a record made of any action taken;
- Finished product control–this requires that the finished product be unadulterated, properly labelled and that the integrity of the finished be protected from the environment.

HACCP and QACCP

All food production activities must be monitored and controlled within the framework of an effective QA programme. The addition of nutrients to a food for the purpose of fortification adds to the control points which have to be considered. Poor manufacturing control leading to excessively high levels of nutrients in the finished product could have important health implications for the consumer if intake of the nutrient reaches the toxic dose. Conversely, low levels of nutrients in the finished product could render it nutritionally ineffective. This could also have serious health implications if the target population in the fortification programme was at high nutritional risk. Poor manufacturing control could also lead to other quality defects related to interactions of added nutrients with other components of the system. The following steps in the implementation of a quality assurance programme in the production of a fortified food have been outlined by Wilson:

- Product specifications–All specifications for fortificants, food vehicle and any other ingredients must be documented as well as acceptable deviations of these. These include specification of particle size, colour, potency, level of fortification as well as any other requirement which might be deemed necessary.

- Product safety assessment–This involves an assessment of microbiological, chemical and physical hazards for all ingredients and the finished product
- Product analysis–Sampling and testing procedures for all ingredients and the finished product must be explicitly stated.
- Determination of critical and quality control points – Based on first hand knowledge of the total process (including the plant facility, equipment and environment) stages at which inadequate control could lead to unacceptable health risk or adversely affect product quality are identified. The system of controls and actions to be taken at each control point are documented.
- Recall system–A mechanism must be put in place whereby product can be recalled if such action becomes necessary.
- QA audit–Periodic checks are necessary to verify that the QA system is effective and product quality is maintained up to the ultimate consumer.
- Feedback mechanism–Response to consumers and other relevant groups to correct any deficiencies discovered.
- Documentation of QA system–Details of the QA programme used in the production of the fortified food must be readily available to relevant individuals and organizations.

Shortcomings of many fortification programmes in the past have been due to failure to establish an adequate quality assurance programme. Evaluation of the fortification of sugar with vitamin A in Guatemala showed that only 30 per cent of samples tested were fortified at levels within the legal limits. A study of iodine content in iodised salt samples obtained from several plants in India also provided an example of the need for greater control in processing. In the determination of critical and other control points for any process accurate flow diagrams outlining the total process have been used. The construction of an accurate flow diagram for any given process requires first hand knowledge of the processing facility and its environs so that all factors which might be expected to impact on product safety could be identified. Recommendations of the FAO/WHO Expert Technical Consultation on "The Use of HACCP in Food Control' included the following:

- Use of HACCP serves to improve food safety control and should be applied on that basis;
- The elabouration of food safety policies by government and international agencies should use risk analysis as the basis for establishing food safety priorities and for focusing inspection

resources. These policies should be implemented through national strategic plans;

- In the post Uruguay Round of GATT, the Codex Alimentarius Commission should recognise the importance of its role in harmonising and establishing food standards, guidelines and recommendations particularly as it relates to safety of food in international trade. Codex should develop a strategic plan which will include a strengthening of the scientific basis for risk analysis, equivalence and the elabouration of its standards, guidelines and other recommendations and should include specific instructions to the Codex Committees on incorporating HACCP.

Analysis of Vitamins and Minerals

Analysis of potency of fortificants and of vitamin and mineral content constitute an important component of the overall analytical requirements in QA/QC programmes for fortification processes. Development or selection of appropriate analytical methodologies must be based on consideration of accuracy and precision of measurements, available facilities and equipment, simplicity of procedure and rapidity of determination.

There are also many other experimental methods which have been developed in various labouratories. The Codex Committee on Methods of Analysis and Sampling is working in close cooperation with ISO, AOAC International and IUPAC to recommend protocols for determining the reliability of analytical test methods and results of labouratory analyses. The community bureau of reference of the Commission of European Communities (BCR) has set up a working group to compare analytical methods used in different labouratories in Europe.

Walter summarised the methods most commonly applied to the analysis of vitamins in foods. HPLC is the technique of choice for many of the vitamins because of the reliability of the method, the rapidity of the determination and the often reduced requirement for rigorous preliminary clean up steps. The drawbacks associated with this technology, however, are high equipment and materials costs and its lack of mobility. A comprehensive review of current methods used in the analysis of vitamins was provided by Lumely.

For analysis of vitamin A content in the MSG fortification programme Muhilal et al. developed a quantitative spectrophotometric method. A semi-quantitative method was also described suitable for field testing in the MSG-vitamin A programme. A rapid quantitative method for the determination of vitamin A content in fortified sugar has been described by Aguilar et al. This is based on the Carr-Price procedure. Adaptations of the method, yielding semi-quantitative results, suitable for field testing have also been reported.

The determination of iodine content in iodised salt has traditionally been carried out using a titrimetric method. Recently a number of HPLC based methods have been used to a large extent. One such method described by de Kleijn was used by the Food Inspection Service in the Netherlands. It utilised reversed phase HPLC with uv detection, and had a detection limit of 2.9 ng KI. Modified HPLC procedures involving precolumn derivatisation have been developed to increase the sensitivity of the analysis.

In this way Verma et al. reported a detection limit of 0.5 ng iodide. Another recent report involved a differential pulse polarographic method which did not require a separation or preconcentration step. For qualitative testing in the field, simple test kits based on the reaction of starch with iodine are available. There are important restrictions to the use of the kits which are currently being used. These are reported to be applicable over a restricted pH range.

Additives commonly used in salt such free-flowing agents or stabilisers of added iodine cause the pH to be elevated above the valid range. In such cases, false negative results are obtained. Modification of this procedure to extend the functional range of testing conditions must be investigated. Users of such field kits need to be educated as to their correct use and their limitations.

In iron fortification programmes, quite often measurement of iron content is inadequate. Bioavailability of the nutrients has been determined by measurement of labelled isotopes absorbed from the diet of human subjects or less complex methods based on animal studies such as the haemoglobin repletion method. Simple chemical methods have been used to approximate bioavailability.

There may be analyses other than the measurement of nutrient content or availability necessitated by a fortification programme. These might include the measurement of colour, particle size or moisture content as well as all testing required for the unfortified food. Based on the identification of critical and other control points, the analytical requirements of the overall quality assurance programme can be determined.

Role of Legislation and Food Control

The primary purposes of food legislation are to protect the health of the consumer and to protect the consumer from fraud. In the case of fortified foods, there is a need to ensure that the population is not at risk of receiving toxic doses of any micronutrient. Food laws must also ensure that the target population does not receive nutritionally ineffective levels of micronutrients. Procedures for monitoring premises where fortified foods are prepared, packed, stored or held for sale as well as mechanisms for penalising defaulters must be clearly defined within the food regulations.

In Switzerland samples of fortified products, both local and imported, are taken from the market annually for analysis of vitamin content by two specialised institutes. If vitamin content is found to be outside of established acceptable limits, the official government agency can disallow sale of the product.

Standards for fortified foods and labelling requirements must also be contained within the food regulations. Standards play an important role in the facilitation of trade, both nationally and internationally. When large differences exist between national standards, however, they can forn technical barriers to trade. In light of the Agreement on Technical Barriers to Trade, the development of international standards for fortified foods is an important step in the elimination of technical barriers.

It has been found that food law is managed most effectively in two parts: a basic food act and food regulations. The act itself should set out broad principles while the regulations should contain the detailed provisions governing the different categories of products. Within the regulations should be found lists of approved fortificant compounds and food standards stating the allowed levels of nutrients in the fortified foods.

This organization gives some flexibility to food law as it is much more difficult to have laws amended than to revise regulations. Prompt revision of regulations may become necessary because of new scientific knowledge, changes in new processing technology or emergencies requiring quick action to protect the public health.

With respect to regulations dealing with fortified foods, changes might be prompted as a result of safety evaluations on nutrient compounds or new information regarding the roles and optimal levels of specific micronutrients in the maintenance of good health. Changes in food processing and packaging technologies could be shown to result in a significant reduction in processing and storage losses of micronutrients, thus requiring a revision in the allowed levels of addition of nutrients. In the face of demonstrated micronutrient deficiencies, regulations regarding standards for certain foods and levels of fortification may need to be revised.

3

Emerging Technologies in Food Preservation

Food Preservation with Chemicals

Some Considerations

- What organisms are the target?
  - bacteria (procaryotes) vs. yeasts or molds (eucaryotes)
  - Gram-negatives vs. Gram-positives
- What are the properties of the food system?
  - pH
  - lipid, protein content
  - natural inhibitors
- How effective is the preservative in food vs. model systems?
- What concentration is needed?
- Is the preservative safe (GRAS)?
- Is it only allowed in certain foods or at certain levels?
- Is it natural or synthetic?

Mode of Action

- In general terms:
  - Killing vs. growth inhibition
  - Specificity for certain organisms or groups
  - Inhibition of spore germination or outgrowth
- More specific effects on cells:
  - Inhibition of nutrient transport
  - Enzyme inhibition
  - Acidification of cell cytoplasm
- Membrane effects
  - Disruption of proton motive force
    a. Undissociated acids transport protons across membrane

   b. Proton released in cytoplasm
   c. Cytoplasm is acidified
 - Alteration of membrane permeability
   a. Uptake inhibited, or leakage of intracellular compounds

Lipophilic Compounds

- Preservatives in this group
  - Benzoic acid and benzoates
  - Parabens
  - Sorbic acid and sorbates
  - Propionic acid and propionates
- Characteristics
  - Can pass through cell membrane
  - More effective at low pH
    a. In undissociated (uncharged) form
    b. Additive effect of low pH
  - Used to control yeasts and molds (bacteria already inhibited)
  - Used in foods susceptible to yeast and mold spoilage
  - Allowable limits ~0.1-0.2%
- Action
  - Inhibit substrate uptake
  - Carry protons into cell
  - Increase energy output of cells

Hydrophilic Compounds

- Preservatives in this group
  - Sulfur dioxide and sulfites
  - Nitrites and nitrates
  - Sodium chloride
  - Sugar
  - Acids
- Characteristics
  - Small molecules
  - Interact with cell components
  - Affected by pH, oxidation/reduction potential

  - May affect available oxygen
  - Affect bacteria as well as yeasts and molds
- Action
  - Sulfites: mechanism unknown
    a. pH of food is lowered
    b. Available oxygen reduced
    c. Essential enzymes inhibited
  - Nitrite: both oxidizing and reducing agent
    a. Key to color development in cured meats
    b. Interferes with key iron-sulfur enzymes
    c. Blocks ATP production from pyruvate
    d. Inhibits vegetative cell growth and spore germination
  - Salt and sugar
    a. Tie up available water
    b. Water lost from cells to equilibrate solute concentration
    c. Mot affected by pH
    d. Salt more inhibitory than sugar
    e. Acids (lactic, acetic, citric, malic, tartaric, succinic)
    f. Lower pH of environment
    g. Acidify cytoplasm if they can enter cell
    h. May chelate ions

Bacteriocins and Antibiotics

- How are these different?
  - Bacteriocins are:
    a. Peptides, proteins, or protein complexes
    b. Synthesized by bacteria
    c. Inhibitory to (usually) closely related bacteria
    d. Membrane-active (disrupt permeability, PMF)
  - Antibiotics are:
    a. Secondary metabolites, not necessarily proteins
    b. Synthesized by many organisms, especially molds and streptomycetes
    c. Inhibitory to a wide range of organisms
    d. Active at various targets in the cell

- Concerns about antibiotics
  - Selection of antibiotic-resistant microorganisms in foods
  - Transfer of resistance genes among organisms
  - Development of resistance to antibiotics used therapeutically
  - Effect of antibiotic residues on humans
- Nisin
  - Only bacteriocin approved for food use in U.S. (processed cheese)
  - Possible adjunct for heat processing
  - Inhibits spore germination
  - Alters membrane permeability
- Bacteriocins as natural food preservatives
  - Produced by GRAS organisms (lactic acid bacteria)
  - Entire culture can be dried down and added to food
  - Mostly affect Gram-positive organisms, at present
  - Rapidly bactericidal
  - Other antimicrobials produced:
    a. Lactic acid
    b. Hydrogen peroxide

Other Chemicals

- Indirect antimicrobials
  - Added for other purposes but have antimicrobial effect
  - A flavoring agents more antifungal than antibacterial
  - Fatty acids and esters probably combine with other chemicals
- Ethylene and propylene oxides
  - Exist as gases and used as fumigants
  - Alkylate reactive groups in proteins, also DNA
  - Sterilizing agent for packaging
- Naturally occurring enzymes and proteins
  - Conalbumin (egg white) binds iron
  - avidin (egg white) binds biotin
  - lysozyme (egg white) degrades Gram + cell walls
  - lactoferrin (milk) binds iron
  - lactoperoxidase (milk) combines with H_2O_2;damages membranes

- Miscellaneous chemicals
  - H_2O_2 sterilizes surfaces, oxidizes
  - Ethanol dries and denatures
  - Polyphosphates chelate divalent ions
  - Wood smoke contains formaldehyde and other inhibitory organic compounds
  - Diethylpyrocarbonate kills yeasts; banned in U.S. because of possible carcinogen formation
- Antimicrobial Chemicals
  - Lipophilic molecules
    a. Sorbic acid CH_3-$(CH)_4$-COOH
    b. Propionic acid CH3CH2COOH
  - Hydrophilic molecules
    a. Sulfur dioxide SO2 sodium nitrite NaNO2
    b. Sulfite SO3= sodium nitrate NaNO3
    c. Bisulfite HSO3- salt NaCl
    d. Metabisulfite S2O5= sugars C6H12O6
    f. Acids RCOOH
  - Bacteriocins and antibiotics
    a. Nisin natamycin
    b. Subtilin tetracycline
  - Other chemicals
    a. Antioxidants: BHA, BHT, TBHQ, PG
    b. Flavoring agents: essential oils, spices
    c. Emulsifiers and surface-active agents: fatty acids and esters
    d. Enzymes and proteins: lactoperoxidase, avidin, lysozyme, lactoferrin
    f. Sterilizing agents: ethylene and propylene oxide, hydrogen peroxide

Preservation by Radiation

Properties of Radiation

- Terms
  - Ionizing radiation (<2000 A) : ionize molecules in its path
  - Curie: quantity of radioactive substance (now becquerel)

- Roentgen: unit of measure for exposure dose of radiation
- Rad: unit of measure for absorption of radiation energy 1 krad = 1000 rads 1 gray = 100 rads
- Ev: energy gained by one electron moving through 1 volt

• Radiation of interest
- UV
  a. Nonionizing, but bactericidal
  b. Absorbed by proteins and nucleic acids → mutations
  c. Poor penetration → limited to surfaces
- Beta rays
  a. Stream of electrons
  b. Produced by linear accelerator
  c. Poor penetration (but better than UV)
- Gamma rays
  a. Emitted from excited nucleus (60Co, 137Cs)
  b. Excellent penetration
- X rays
  a. Similar to Gamma rays
  b. High-speed electrons bombard heavy metal
- Microwaves
  a. Long wavelengths → nonionizing
  b. Rapidly alternating field
  c. Heat generated by molecular friction of oscillations

Radiation vs. Microorganisms

• Factors controlling effectiveness
- Types of organisms
  a. Gram-positive more resistant than Gram-negative
  b. Sporeformers more resistant than nonspore-formers
  c. Yeasts and molds more resistant than bacteria
  d. Extremely resistant organisms exist
- Number of organisms: larger number → less effective
- Age of organisms: mostresistant in lag phase
- Composition of medium: proteins are protective
- Oxygen level: higher resistance in anaerobic environment

- Physical state:higher resistance in dry or frozen state
- Why are there such differences in resistance?
  - Consider effects of radiation:
    a. Rapidly alternating field
    b. Generate ionized and electrically excited molecules
    c. Radiolysis (splitting) of water molecules
    d. Free radical and peroxide formation
    e. Damage to DNA
  - Organisms can vary in:
    a. Cell wall and cell membrane composition
    b. Enzyme systems to quench free radicals, peroxides
    c. Content of reducing compounds
    d. DNA repair systems
  - Rsafety in numbers phenomenon
  - Nongrowing cells are not replicating DNA
  - Availability of compounds to quench free radicals
  - Availability of oxygen for peroxide formation
  - Less water available in dry or frozen state
- Extremely radiation-resistant organisms:
  - Deinococcus
  - Rubrobacter
  - Deinobacter
  - Acinetobacter
  - Viruses (small targets) spores (esp. C. botulinum)

Radiation of Foods

- Sources:
  - Gamma rays from Cs or Co source
  - Electron beam from linear accelerator
- Treatment levels
  - Radappertization: * commercial sterilization - 30-40 kGy * harshest treatment → chem. changes
  - Radicidation: * pasteurization - 2.5-10 kGy * kill Salmonella in raw poultry * treat spices
  - Radurization: * "mild" pasteurization - 0.75-2.5 kGy * reduce

spoilage organisms * extend shelf life of produce * kill insect eggs, larvae, etc.

- Effects in Foods
  - Chemical changes with higher levels of irradiation:
    a. Off-flavors from free-radical, peroxide reactions
    b. Lipid oxidat ion
    c. Release of NH3, H2S from proteins and nitrogenous cpds
    d. Partial destruction of B-vitamins
    e. Pectin and cellulose degradation → softening
    f. Enzymatic changes
  - Reduce effects by:
    a. Reduce temperature → slow free radical movement
    b. Reduce oxygen → fewer oxidative free radicals
    c. Add free radical scavengers
    d. Remove off-flavors
- Foods approved for irradiation by U.S. FDA:

Food	Purpose	Dose Limit
fruits and vegetables	slow growth and ripening	up to 1 kGy control insects
dry herbs, spices seeds, teas, seasonings	kill insects control microorganisms	up to 30 kGy 0.3 kGy - 1 kGy
pork	control Trichinella spiralis	
white potatoes	inhibit sprout development	50 - 150 Gray
wheat, wheat flour	control insects	200 - 500 Gray

Low Temperature Processes

Basic Principle

- Slow down or stop activities of microorganisms
- Not a lethal process (but some viability loss may occur)
- Q10 = ratio of reaction velocities at temperatures 10°C apart
  - For most enzyme-catalyzed reactions, Q10 = 1.5-2.5
  - Reaction rates decrease by ~2x with each 10° fall in temp.

Temperature Ranges

- Chilling = 10-15°C - fruit and vegetable storage
- Refrigeration = 0-7°C - for perishable and semiperishable foods
- Freezing = below -18°C - can still get microbial growth

Physical Effects at Low Temperature

- Relative humidity generally higher at chill temperatures
- Reduced aw as available water freezes
- Oxygen solubility increases with decreasing temperature
- Some pH changes may occur

Important Organisms

- Psychrophiles and psychrotrophs can grow at low temp.
- Low-temperature spoilage organisms mostly Gram-negatives:
  - Pseudomon as Acinetobacter
  - Alcaligenes Flavobacterium
- Many yeasts and molds can grow at low temp
- Most pathogens don't grow below 6C, but exceptions occur:
  - Listeria monocytogenes
  - Vibrio parahaemolyticus
  - Aeromonas hydrophila

Freezing

- First blanch in hot water or steam to:
  - Inactivate enzymes
  - Reduce microorganisms
  - Fix color
  - Wilt leafy vegetables
  - Displace entrapped air
- Quick (-20°C in 30 min) vs. slow (-20°C in 3 to 72 hr) freezing:
  - Small vs. large ice crystals
  - Time of exposure to concentrated solutes
  - Opportunity to adapt % halted vs. continued metabolism

Effects on Microorganisms

- Population level
  - some cells killed during freezing process

  - number of survivors will decline upon storage
  - viability loss is greatest close to freezing point
  - don't expect complete lethality
- Cellular level
  - Free water freezes; bound water doesn't
  - Cells dehydrated as free water reduced
  - Solutes concentrated in remaining available water
  - Viscosity of cytoplasm increased
  - Gases (O_2 and CO_2) lost → aerobic respiration reduced
  - pH changes in cytoplasm
  - Proteins denatured
  - Phase transition causes membrane damage
- Slower changes during thawing generally more harmful

Characteristics of Psychrotrophs

- Most important spoilage organisms are aerobes or facultative
- Membrane lipids fluid at low temperature - unsaturated fatty acids
- Extracellular polysaccharides produced - cryoprotectants?
- Enzymes remain active at low temperature
- Enzymes relatively heat-labile

High-Temperature Processes

Types of Processes

- Pasteurization
  - Destroy pathogens; destroy or reduce spoilage organisms
  - Low temp-long hold: 63°C, 30 min.
  - High temp-short time: 72°C, 15 sec.
  - Kill Mycobacterium tuberculosis and Coxiella burnetii
  - Thermodurics and thermophiles survive
- Sterilization
  - Destroy all viable organisms
  - Commercial sterility: only significant organisms destroyed
- Ultra-high temperatures
  - 140-150°C for a few seconds gives commercial sterility
  - Continuous process + aseptic packaging → shelf-stable milk

Factors Affecting Heat Resistance

- Moisture
  - Dry cells more resistant
  - Water may facilitate protein denaturation from heat
- Fat content: higher resistance in presence of fat (lower aw?)
- Salts: some protect (lower aw?), others sensitize
- Carbohydrates: sugars lower aw and increase resistance
- pH: highest resistance at optimum pH (acid foods need less heat)
- Proteins: protective effect
- Number of organisms: higher number → higher resistance
- Age of organisms: log phase most sensitive; stationary resistant
- Growth temperature: resistance increases with increasing temp.
- Inhibitory compounds: less resistance if inhibitors present
- Time and temp: greater killing at higher temp, longer time

Microbial Heat Resistances

- Resistance related to optimum growth temperature:
  - Psychrophiles < mesophiles < thermophiles
  - Function of heat sensitivity of enzymes
- Gram + more resistant than Gram -
- Yeasts and molds (including spores) fairly sensitive
- Sporeformers (and their spores) most resistant
  - Protoplast is dehydrated
  - Concentration of ions, polymers (e.g. Ca-dipicolinate)
  - At high growth temperature spore moisture is reduced

Thermal Destruction Concepts

- Target: bacterial endospores, esp. Clostridium botulinum
- Thermal death time: time to kill a given # of org. at a given temp.
- D value:
  - Decimal reduction time
  - Time to destroy 90% of organisms
  - Reciprocal of slope of survivor curve
  - Compare relative heat resistances by comparing D values
- Z value:

  - °F to decrease D by 90%
  - Reciprocal of slope of D vs. oF
  - Can calculate process times at other temperatures
- F value: equivalent time in min. at 250°F of a heat process
- 12 D concept:
  - Process lethality standard in the canning industry
  - Reduce C. botulinum population by 10-12
  - Reduce probability of survival of spores to 10-12
  - For low-acid foods (pH 4.6 and above)

High-heat processe

- Conventional canning
  - Nonsterile food placed in nonsterile metal or glass container
  - Heat process in closed container
- Aseptic packaging
  - Sterile food placed in sterile containers, sealed aseptically
  - Food pumped through heat exchanger for rapid heating
  - Packaging materials sterilized by hydrogen peroxide

Characteristics of thermophiles

- Organisms
  - Growth temps: min. 45°C; opt. 50-60°C; max. >70°C
  - Many free-living types
  - Bacilli and clostridia most important in foods
- Growth
  - Short lag time
  - Short log phase
  - Short generation time
  - Often rapid die-off once stationary phase reached
  - May not survive exposure to lower temperatures
- Enzymes
  - High heat resistance → must maintain stable conformation
  - Presence of substrate may increase heat resistance
  - May have high hydrophobic amino acid content
  - May form intramolecular bonds

  - May complex with ions
- Ribosomes
  - More heat-stable than those of mesophiles
  - Higher G-C content → more hydrogen bonding
- Nutrient requirements
  - May need more nutrients at higher temperature
  - Less efficient metabolism?
  - Some enzymes inactivated
- Oxygen demand
  - Faster growth rate requires more oxygen
  - Lower oxygen solubility at high temperature
- Lipids and membranes
  - Higher degree of saturation at higher temperature
  - More hydrophobic bonding with increased saturation
  - Membrane stability may be key factor in heat damage

Drying

Types of Foods

- Dry, desiccated: aw = 0-0.6; <25% moisture
  - Combination of dry conditions + heat required
  - Sun drying: raisins
  - Spray and drum drying:
    a. Nonfat dry milk powder
    b. Evaporated milk (40% of water left)
- Freeze-dried: same moisture as dry food
  - Freeze product, then sublime water with vacuum (+ heat)
  - Rule of thumb moisture level: 2
- Intermediate moisture: aw = 0.6-0.85; 15-50% moisture
  - Remove water, or add ingredients to tie it up
  - Traditional foods: dried fruits, jams, fermented sausages
  - Newer foods: cat and dog foods, shelf-stable meat products
  - Processes
    a. Adsorption method
    b. First dry food

c. Rehumidify to adjust aw
d. Desorption method
e. Place food in solution of higher osmotic pressure
f. Allow equilibration
g. These reach same aw but not necessarily same moisture
h. Adsorption method holds less water, is more inhibitory
i. Add humectants (glycerol, sucrose) to tie up water

Effects on Microorganisms

- Preliminary treatments (blanching, alkali, SO_2) reduce load
- Moisture is reduced below that required for growth
  - Bacteria won't grow (Gram- more sensitive than Gram+)
  - Some osmophilic yeasts may grow
  - Molds are primary spoilage organism
- Drying may not kill (unless heat applied) but can injure cells
  - Damage results from ice crystal formation (in freeze-drying)
  - Stress of high solute concentrations in remaining water
- Preservative effect: dry cells less sensitive to heat, radiation
- Expect sporeformers as contaminants of dried foods
- Pathogens of concern: sporeformers, Staph. aureus
- Freeze-drying is popular preservation method for microbial cultures

Fermentation

What is Fermentation?

- Carbohydrates and related compounds oxidized to yield energy
- Final electron acceptors are organic compounds derived from carbos
- Only partial oxidation achieved; not all available energy released
- Redox balance of substrate vs. products
- Chemical change brought about in organic substrate by action of microbial enzymes

Features

- Microorganisms responsible for characteristics of food
- Shelf life is usually longer than that of raw product

- Preservation by acid, ethanol, other antimicrobials
- Aroma and flavor due to fermenting organism
- Vitamin content and digestibility can be increased

Types of Fermentations

- Acid food + free sugar: yeasts make alcohol (wines)
- Less acid + free sugar: bacteria make acids (cheese)
- Little simple sugar: require enzymes to break down complex carbohydrates and proteins
  - Malt (germinating barley) in beer
  - Koji (Aspergillus oryzae culture) in soy products (miso)

Products and Processes

- Glucose → lactic acid (homofermentative)
  - Embden-Meyerhof-Parnas (EMP) pathway to pyruvate
- Glucose → acid + CO2 + ethanol (heterofermentative)
  - Hexose monophosphate or pentose path to pyruvate
- Glucose → pyruvate → ethanol (yeasts) → acetic acid (Acetobacter)
- Glucose or lactate → propionate + acetate + CO2 (propionibacteria)

Important Organisms

- Lactic acid bacteria
  - Lactobacillu
  - Carnobacterium
  - Leuconostoc
  - Enterococcus
  - Pediococcus
  - Lactococcus
  - Streptococcus
  - Vagococcus
- Yeasts
  - Saccharomyces sp. (esp. S. cerevisiae)
  - Zygosaccharomyces Candida
- Molds
  - Aspergillus Penicillium Geotrichum Rhizopus

Typical Fermentation Process

- Substrate disappears as cell mass increases
- Primary metabolic products (acids) accumulate during growth
- pH drops if acids produced
- Growth and product formation stop as substrate is depleted
- Secondary products (if any) accumulate after growth stops
- For multiple substrates, easiest to metabolize disappears first
- Microbial succession depends on substrate and acid levels
  - Bacteria, then yeasts, then molds grow
  - Sugar, then other small molecules, then polymers used

Food Products

- From milk:
  - Cheese, yogurt, sour cream, buttermilk
  - lactic acid bacteria (lactobacilli, streptococci) start ferm.
  - special organisms for unique characteristics
- Meats:
  - fermented sausages, hams, fish (Asia)
  - lactic acid bacteria (lactobacilli, pediococci), molds
- From plants:
  - lactic fermentation (pickles, olives) by lactic acid bacteria
  - coffee, cocoa beans fermented by yeasts, bacteriaA
  - soy products (soy sauce, tempeh, miso) by bacteria, molds
- Beverages:
  - beer and ale (yeasts make ethanol from cereal carbos)
  - wines (ethanol fermentation from grapes, other fruits)
  - distilled spirits (must distill to concentrate ethanol)
  - vinegar (ethanol oxidized to acetic acid)
- Breads:
  - sourdough (yeast + lactobacilli)
  - crackers, raised breads (yeasts)
- Single cell protein:
  - how cheaply and efficiently can cells be grown?
    a. Waste materials as substrate (bacteria, yeast, molds)

   b. Sunlight and CO_2 (algae)
- uses in animal feeds (frequently) or human foods
- prefer protein to whole cells
   a. High nucleic acids → kidney stones, gout

Benefits of Fermented Foods and Fermentative Organisms

- Reduce lactose intolerance
  - Less lactose available
  - Bacterial enzymes available % lower serum cholesterol
  - Reduced synthesis by the body
  - Cholesterol assimilated by microorganisms
- Anticancer effects
  - Lower activity of enzymes producing carcinogens
  - Microbial metabolism of procarcinogens.

4

Fruit and Vegetable Processing

Importance of fruit and vegetables

In developing countries agriculture is the mainstay of the economy. As such, it should be no surprise that agricultural industries and related activities can account for a considerable proportion of their output. Of the various types of activities that can be termed as agriculturally based, fruit and vegetable processing are among the most important.

Both established and planned fruit and vegetable processing projects aim at solving a very clearly identified development problem. This is that due to insufficient demand, weak infrastructure, poor transportation and perishable nature of the crops, the grower sustains substantial losses. During the post-harvest glut, the loss is considerable and often some of the produce has to be fed to animals or allowed to rot.

Even established fruit and vegetable canning factories or small/ medium scale processing centres suffer huge loss due to erratic supplies. The grower may like to sell his produce in the open market directly to the consumer, or the produce may not be of high enough quality to process even though it might be good enough for the table. This means that processing capacities will be seriously underexploited.

The main objective of fruit and vegetable processing is to supply wholesome, safe, nutritious and acceptable food to consumers throughout the year. Fruit and vegetable processing projects also aim to replace imported products like squash, yams, tomato sauces, pickles, etc., besides earning foreign exchange by exporting finished or semi-processed products.

The fruit and vegetable processing activities have been set up, or have to be established in developing countries for one or other of the following reasons:

- Diversification of the economy, in order to reduce present dependence on one export commodity;
- Government industrialisation policy;
- Reduction of imports and meeting export demands;
- Stimulate agricultural production by obtaining marketable products;

- Generate both rural and urban employment;
- Reduce fruit and vegetable losses;
- Improve farmers' nutrition by allowing them to consume their own processed fruit and vegetables during the off-season;
- Generate new sources of income for farmers/artisans;
- Develop new value-added products.

Heat preservation/heat processing

Various Degrees of Preservation

There are various degrees of preservation by heating; a few terms have to be identified and understood.

- Sterilisation. By sterilisation we mean complete destruction of micro-organisms. Because of the resistance of certain bacterial spores to heat, this frequently means a treatment of at least 121° C (250° F) of wet heat for 15 minutes or its equivalent. It also means that every particle of the food must receive this heat treatment. If a can of food is to be sterilised, then immersing it into a 121° C pressure cooker or retort for the 15 minutes will not be sufficient because of relatively slow rate of heat transfer through the food in the can to the most distant point.
- "Commercially sterile". Term describes the condition that exists in most of canned or bottled products manufactured under Good Manufacturing Practices procedures and methods; these products generally have a shelf-life of two years or more.
- Pasteurized means a comparatively low order of heat treatment, generally at a temperature below the boiling point of water. The more general objective of pasteurization is to extend product shelf-life from a microbial and enzymatic point of view; this is the objective when fruit or vegetable juices and certain other foods are pasteurized.
- Pasteurization is frequently combined with another means of preservation - concentration, chemical, acidification, etc.
- Blanching is a type of pasteurization usually applied to vegetables mainly to inactivate natural food enzymes. Depending on its severity, blanching will also destroy some microorganisms.

Determining Heat Treatment/Thermal Processing Steps

Since heat sufficient to destroy micro-organisms and food enzymes also usually has adverse effects on other properties of foods, in practice the minimum possible heat treatment should be used which can guarantee freedom from pathogens and toxins and give the desired storage life; these aims will determine the choice of heat treatment. In order to safely preserve foods using heat treatment, the following must be known:

- What time-temperature combination is required to inactivate the most heat resistant pathogens and spoilage organisms in one particular food?
- What are the heat penetration characteristics in one particular food, including the can or container of choice if it is packaged?

Preservation processes must provide the heat treatment which will ensure that the remotest particle of food in a batch or within a container will reach a sufficient temperature, for a sufficient time, to inactivate both the most resistant pathogen and the most resistant spoilage organisms if it is to achieve sterility or "commercial sterility", and to inactivate the most heat resistant pathogen if pasteurization for public health purposes is the goal.

Different foods will support growth of different pathogens and different spoilage organisms so the target will vary depending upon the food to be heated.

Technological principles of pasteurization

Physical and chemical factors which influence pasteurization process are the following:

- Temperature and time;
- Acidity of the products;
- Air remaining in containers.

Pasteurization processes. In pasteurising certain acid juices for example, there are two categories of processes:

- Low pasteurization where pasteurization time is in the order of minutes and related to the temperature used; two typical temperature/time combinations are as following:
- 63° C to 65° C over 30 minutes, or
- 75° C over 8 to 10 minutes.

Pasteurization temperature and time will vary according to:

- Nature of product; initial degree of contamination;
- Pasteurized product storage conditions and shelf life required.

In this first category of pasteurization processes it is possible to define three phases:

- Heating to a fixed temperature;

- Maintaining this temperature over the established time period (= pasteurization time);
- Cooling the pasteurized products: natural (slow) or forced cooling.
- Rapid, high or flash pasteurization is characterized by a pasteurization time in the order of seconds and temperatures of about 85° to 90° C or more, depending on holding time. Typical temperature/time combinations are as follows:
- 88° C (190° F) for 1 minute;
- 100° C for 12 seconds;
- 121°C for 2 seconds.

While bacterial destruction is very nearly equivalent in low and in high pasteurization processes, the 121° C/2 seconds treatment give the best quality products in respect of flavour and vitamin retention. Such short holding times, however, require special equipment which is more difficult to design and generally is more expensive than the 63-65 ° C/30 minutes type of processing equipment.

In flash pasteurization the product is heated up rapidly to pasteurization temperature, maintained at this temperature for the required time, then rapidly cooled down to the temperature for filling, which will be performed in aseptic conditions in sterile receptacles. Taking into account the short time and rapid performance of this operation, flash pasteurization can only be achieved in continuous process, using heat exchangers.

Industrial applications of pasteurization process are mainly used as a means of preservation for fruits and vegetable juices and specially for tomato juice.

Thermopenetration

The thermopenetration problem is extremely important, especially in the case of the pasteurization of products packed in glass containers because it is the determining factor for the success of the whole operation.

During pasteurization it is necessary that a sufficient heat quantity is transferred through the receptacle walls; this is in order that the product temperature rises sufficiently to be lethal to micro-organisms throughout the product mass. The most suitable and practical method to speed up thermopenetration is the movement of receptacles during the pasteurization process.

Rapid rotation of receptacles around their axis is an efficient means to accelerate heat transfer, because this has the effect, among others of rapidly mixing the contents. The critical speed of for this movement is generally about 70 rotations per minute (RPM). This enables a more uniform heating of products, reducing heating time and organoleptic degradation. Heating may precede or follow packaging. These principles of different temperature time combinations very largely determine the design parametres for heat preservation equipment and commercial practices.

The food processor will employ no less than that heat treatment which gives the necessary degree of micro-organism destruction. This is further ensured by periodic inspection by local sanitary authorities or by the importing countries sanitary services. However, the food processor also will want to use the mildest effective heat treatment to ensure highest food quality.

It is convenient to separate heat preservation practices into two broad categories: one involves heating of foods in their final containers, the other employs heat prior to packaging. The latter category includes methods that are inherently less damaging to food quality, where the food can be readily subdivided (such as liquids) for rapid heat exchange. However, these methods then require packaging under aseptic or nearly aseptic conditions to prevent or at least minimise recontamination. On the other hand, heating within the package frequently is less costly and produces quite acceptable quality with the majority of foods and most of our present canned food supply is heated in the package. In practice, therefore, most of the canned food produced locally in developing countries should be heated within the package.

Banana and plantain processing technologies

Traditional Processing

Products: Uses and Dietary Significance

Most of the world's bananas are eaten either raw, in the ripe state, or as a cooked vegetable, and only a very small proportion are processed in order to obtain a storable product. This is true both at a traditional village level with both dessert and cooking bananas and when considering the international trade in dessert bananas.

In general, preserved products do not contribute significantly to the diet; however, in some localised areas the products are important in periods when food are scarce. Probably the most widespread and important product is flour preparation from unripe banana and plantains by sun-drying. In Uganda, dried slices known as "mutere" are prepared for storage from green bananas, the dried slices being either used directly for cooking or after grinding into a flour.

"Mutere" is used chiefly as a famine reserve and does not feature largely in the diet under normal conditions. In Gabon, plantains are sometimes made into dried slices which can be stored and used on long journeys, and plantains are used in Cameroon to prepare dried pieces which are stored and ground as needed into flour for use in cooking a paste known as "fufu". Dried green banana slices are also used in parts of South and Central America and West Indies for preparing flour.

The other nutritionally important product is beer which is a major product in Uganda, Rwanda and Burundi where green banana utilisation is particularly high.

Preservation Methods and Processes

Drying

Both ripe and unripe bananas and plantains are normally peeled and sliced before drying, although banana figs are sometimes prepared from whole ripe fruit. Sun drying is the most widespread technique where the climate is suitable but drying in ovens or over fires is also practiced.

In west Africa, plantains are often soaked and sometimes parboiled before drying. The slices of unripe fruit are normally spread out on bamboo frameworks; or a cemented area; or on a mat; or on a swept-bare patch of earth; or on a roof; or sometimes on stones outcrops or sheets of corrugated iron.

Oven-drying of ripe bananas is practiced in Polynesia as a mean of preserving the fruits, which are then wrapped in leaves and bound tightly to store until needed. In East Africa a method has been reported that involves drying the peeled bananas on a frame placed over a fire for 24 hr before drying in the sun, to accelerate the process.

Product Stability and Storage Problems

There is little experimental data on the storage life of the traditionally made banana and plantain products.

Potential for Scaling up of Traditional Processes to Industrial Level

Many banana products are now produced on an industrial scale, including the traditional banana figs and flour, and the processing techniques are described below. One of the main problems encountered has been the susceptibility of banana products to flavour deterioration and discolouration and in the past many products reaching the market have been of poor quality. A great deal of research has been directed to overcoming these problems, although however good the resultant products are they cannot compare in flavour and other characteristics with the fresh banana fruit. Indeed, an important constraint on the large-scale development of banana processing is the lack of demand for banana products since the fresh fruit is available throughout the year in most parts of the tropical world. The production of beer from banana and plantains has not been scaled up to an industrial level, and while an important product in localised areas of tropical Africa, the market is rapidly declining in favour of European-type brews produced locally.

Industrial Processing

Products and Uses

The main commercial products made from bananas are canned or frozen purée, dried figs, banana powder, flour, flakes, chips (crisps), canned slices and jams. Banana products can be divided roughly into two types - those for direct consumption, such as figs, and those for use in food manufacturing industry, for example purées and powder. Banana figs, or fingers as they are sometimes known, are usually whole, peeled fruit carefully dried so as to retain their shape, although sometimes the fruit is sliced or halved to facilitate drying. Banana and plantain chips (crisps) are thinly sliced pieces of fruit fried in oil and eaten as a snack like potato chips (crisps). The main use of canned slices is in tropical fruit salads. Banana flakes are used as a flavouring or in breakfast cereals. Banana purée find use mainly in the production of baby foods. Banana flour is said to be highly digestible and is used in baby and invalid foods, but can also be used in the preparation of bread and beverages. Banana powder is used chiefly in the baking industry for the preparation and fillings for cakes and biscuits and is also used for invalid and baby foods.

Processing Technology

In general, to obtain a good-quality product from ripe-bananas the fruit is harvested green and ripened artificially under controlled conditions at the processing factory. After ripening, the banana hands are washed to remove dirt and any spray residues, and peeled. Peeling is almost always done by hand using stainless steel knives, although a mechanical peeler for ripe bananas has been developed, capable of peeling 450 Kg of fruit per hour (Banana Bulletin, 1974).

The peeling of unripe bananas and plantains is facilitated by immersing the fruit in hot water. For example, immersion in water at 70-75 ° C for 5 min. has been suggested as an aid for peeling green bananas for flour production, while the peeling of green bananas for freezing has been facilitated by immersion in water at 93° C for 30 min.

Banana Figs

Fully ripe fruits with a sugar content of about 19.5per cent are used and are treated with sulphurous acid after peeling, then dried as soon as possible afterwards. Various drying systems have been described using temperatures between 50 and 82° C for 10 to 24 hr to give a moisture content ranging from 8 to 18per cent and a yield of dried figs of 12 to 17per cent of the fresh banana on the stem.

One factory in Australia uses a solar heat collector on the roof to augment the heat used for drying bananas. Bananas can also be dried by osmotic dehydration, using a technique which involves drawing water from 1/4-in. thick banana by placing them in a sugar solution of 67 to 70 deg. Brix for 8 to 10 hr. followed by vacuum-drying at 65 to 70° C, at a vacuum of 10 mm Hg for 5 hr. The moisture content of the final products is 2.5per cent or less, much lower than that achieved by other methods.

Banana Puree

Banana purée is obtained by pulping peeled, ripe bananas and then preserving the pulp by one of three methods: canning aseptically, acidification followed by normal canning, or quick-freezing. The bulk of the world's purée is processed by the aseptic canning technique. Peeled, ripe fruits are conveyed to a pump which forces them through a plate with 1/4-in. holes, then onto a homogeniser, followed by a centrifugal de-aerator, and into a receiving tank with 29in. vacuum, where the removal of air helps prevent discolouration by oxidation.

The purée is then passed through a series of scraped surface heat exchangers where it is sterilised by steam, partially cooled, and finally brought to filling temperature. The sterilised purée is then packed aseptically into steam-sterilised cans which are closed in a steam atmosphere.

Banana Slices

Several methods for canning of banana slices in syrup are used. Best-quality slices are obtained from fruit at an early stage of ripeness. The slices are processed in a syrup of 25 deg. Brix with pH about 4.2, and in some processes calcium chloride (0.2per cent) or calcium lactate (0.5per cent) are added as firming agents.

A method for producing an intermediate-moisture banana product for sale in flexible laminate pouches has been developed. Banana slices are blanched and equilibrated in a solution containing glycerol (42.5per cent), sucrose (14.85per cent), potassium sorbate (0.45per cent), and potassium metabisulphite (0.2per cent) at 90 deg. C for 3 min. to give a moisture content of 30.2per cent.

Banana Powder

In the manufacture of banana powder, fully ripe banana pulp is converted into a paste by passing through a chopper followed by a colloid mill. A 1 or 2 per cent sodium metabisulphite solution is added to improve the colour of the final product. Spray- or drum-drying may be used, the latter being favoured as all the solids are recovered. A typical spray dryer can produce 70 kg powder per hour to give yields of 8 to 11per cent of the fresh fruit, while drum-drying gives a final yield of about 13per cent of the fresh fruit. In the latter method the moisture content is reduced to 8 to 12 per cent and then further decreased to 2 per cent by drying in a tunnel or cabinet dryer at 60° C.

Banana Flour

Production of flour has been carried out by peeling and slicing green fruit, exposure to sulphur dioxide gas, then drying in a counter-current tunnel dryer for 7 to 8 hr. with an inlet temperature of 75° C and outlet temperature of 45° C, to a moisture content of 8per cent, and finally milling.

Banana Chips (Crisps)

Typically, unripe peeled bananas are thinly sliced, immersed in a sodium or potassium metabisulphite solution, fried in hydrogenated oil at 180 to 200° C, and dusted with salt and an antioxidant. Alternatively, slices may be dried before frying and the antioxidant and salt added with the oil. Similar processes for producing plantain chips have been developed.

Banana Beverages

In a typical process, peeled ripe fruit is cut into pieces, blanched for 2 min. in steam, pulped and pectolytic enzyme added at a concentration of 2 g enzyme per 1 kg pulp, then held at 60 to 65° C and 2.7 to 5.5 pH for 30 min. In a simpler method, lime is used to eliminate the pectin. Calcium oxide (0.5per cent) is added to the pulp and after standing for 15 min. this is neutralised giving a yield of up to 88per cent of a clear, attractive juice. In another process banana pulp is acidified, and steam-blanched in a 28-in Hg vacuum which ensures disintegration and enzyme inactivation. The pulp is then conveyed to a screw press, the resulting purée diluted in the ratio 1:3 with water, and the pH adjusted by further addition of citric acid to 4.2 to 4.3, which yields an attractive drink when this is centrifuged and sweetened.

Jam

A small amount of jam is made commercially by boiling equal quantities of fruit and sugar together with water and lemon juice, lime juice or citric acid, until setting point is reached.

Product Stability and Spoilage Problems

All dried banana products are very hydroscopic and susceptible to flavour deterioration and discolouration, but this can be overcome to some extent by storing in moisture-proof containers and sulphiting the fruit before drying to inactivate the oxidases.

The dried products are also liable to attack by insects and moulds if not stored in dry conditions, although disinfestation after drying by heating for 1 hr to 80° C or by fumigation with methyl bromide ensures protection against attack. Banana powder is said to be stored for up to a year commercially and flakes have been stored in vacuum-sealed cans with no deterioration in moisture, colour or flavour for 12 months.

Banana chips tend to have a poor storage life and to become soft and rancid. However, chips treated with an antioxidant have been stored satisfactorily at room temperature in hermetically sealed containers up to 6 months with no development of rancidity.

Quality Control Methods

In general a good quality product is obtained if fruit is harvested at the correct stage of maturity and, where appropriate, ripened under controlled conditions. For example, in the case of banana figs, the fruit should be fully mature (sugar content of 19.5per cent or above) or the final product is liable to be tough and lacking in flavour. However, if over-ripe fruit is used, the figs tend to be sticky and dark in colour, so the fruit must be fully yellow but still firm.

For banana flour, which is prepared from unripe bananas, the fruit is harvested at three-quarters the full-ripe stage and is processed within 24 hr. prior to the onset of ripening. If less mature fruit is used, the flour tastes slightly astringent and bitter due to the tannin content. Bananas harvested between 85 and 95 days after the emergence of the inflorescence, with a pulp-to-peel ratio of about 1.7, were considered to be most suitable for the deep-fat frying.

Other criteria suggested for assessing maturity were beta-carotene and reducing sugar content, both of which increase with increasing maturity and pH which decreases as the fruit ripens, and these should be, respectively, about 2000 μg/100 g, less than 1.5per cent and 5.8 or above. Browning was found to occur if the sugar content was higher than 1.5per cent. The determination of crude fat in processed chips is also considered to be a necessary quality control measure. It is important to remove all impurities prior to processing of products, and this is done by washing to remove dirt and spray residues and control on the processing line so that substandard fruit can be removed.

Preparation Methods for Fresh Bananas and Plantains

The main ways of preparing fresh bananas for consumption are boiling or steaming, roasting or baking and frying. Boiling followed by pounding into "fufu" is also widely adopted in certain areas of the tropics.

Boiling or Steaming

Plantains and bananas are often prepared simply by boiling in water, either in their peel or after peeling, and either ripe or unripe; if unripe, the fruit is scraped thoroughly after peeling to remove all traces of fibrous material. The boiled fruit is eaten alone or more usually accompanied by a sauce. This preparation technique is widely used in West Africa.

Roasting or Baking

Unpeeled or peeled fruit, either ripe or unripe, is roasted simply by placing in the ashes of a fire or in an oven. This method is widely used in West Africa, East Africa and the South Pacific islands. For example, ripe plantains are placed unpeeled in an oven and when partly brown and tender, removed and peeled, then replaced in the oven and roasted evenly.

Frying

Ripe or unripe plantains or bananas are often peeled, sliced and cooked in oil, particularly in West Africa and in parts of South America and the West Indies. Similar products are also made in East Africa. Typically, ripe plantains are peeled, cut into slices or split lengthways, and fried in palm oil or with groundnut oil, the pieces being served either hot with a sauce or with fried eggs, or cold as a snack.

Pounding

Pounding is a process, used particularly in West Africa, for preparing most perishable staple food crops including plantains, cassava, yams and cocoyams to obtain a paste or dough known as "fufu" (also spelled "foofoo", "foutou", "foufou"). The plantains are peeled or boiled and peeled after boiling and pounded in a wooden mortar, the resulting paste normally being eaten with soup or a spiced sauce of meat and vegetables, but sometimes after wrapping in leaves and steaming.

Banana and its Market

The present study is directed neither towards justification nor criticism of the United Fruit Company's operations over the more than half century of its history. We feel that it would be an essentially sterile exercise to attempt to disentangle the skein of events since 1899. Since then, a banana industry of sufficient size to assume importance in international commerce literally has been created by imaginative traders, with the United Fruit Company exercising a dominant role in the process. To reconstruct and appraise the historical record in a way that would allow recriminations to be balanced against solid accomplishments would call for an omniscient judgement that we do not feel we possess.

Rather, we have set ourselves the much more modest, though still not unambitious, task of attempting to contribute towards an understanding of what the banana industry is, of who benefits from it and by how much, and to appraise the record of the United Fruit Company in the industry as it operates today. Our perspective, then, is contemporary rather than historic. Most of our field work was completed in 1956, and our major concentration is upon operations in the year 1955, the latest year for which national accounts data for the several banana-producing countries were available in relatively complete form.

We are acutely conscious of the fact that our study deals with a field in which the ideas of most people are coloured by strongly held preconceptions based upon an interpretation of past events rather than upon an examination of the current record. It has seemed to us that a useful service could be performed by describing the present organization and procedure of banana production and marketing, by measuring what can be measured, and by limiting subjective judgements to matters not susceptible to appraisal in objective terms. Where we are forced to make value judgements, it is our hope that at least relative objectivity may be achieved through the circumstance of a joint authorship that combines North American with Latin American outlooks. More particularly, our interest in the United Fruit Company record is focused upon its impact upon the economic development process in the six Latin American republics selected for intensive study-Guatemala, Honduras, Costa Rica, Panama, Colombia, and Ecuador. These were singled out because their combined banana shipments represent close to 60 percent of the tonnage weight of world banana exports, and because about 95 percent of the bananas handled by the United Fruit Company in 1955 was produced in or purchased from these six sources. Through this sampling procedure, we could limit our field work to supportable dimension, and still cover the bulk of the world banana production for export, and a preponderant portion of United's banana procurement operations.

In concentrating upon United Fruit's contributions to economic development in these six countries, we are guided by our conviction that the widest possible diffusion of vigorous economic growth is one of the most important concerns of the contemporary free world. We are convinced that adherence to and strengthening of democratic institutions in the less developed countries of the world depends in large measure upon the demonstration that aspirations for general economic progress can be realized under free institutions. And we believe that the flow of investment capital from the capital-generating nations of high industrial development to the capital-poor countries is a major instrument for helping to stimulate balanced growth in the latter.

We start with the premise that investment capital, private or public, will not continue to flow unless it receives a return judged to be adequate. Therefore we shall examine the profitability of United Fruit investments in the six republics upon this criterion. But it seems equally clear to us that continuing hospitality for foreign private investment ventures cannot be expected to endure unless there is clear evidence that it is contributing to the development of the host nations to a degree that would not be realizable without it.

Accordingly, we shall examine the United Fruit record to see whether or not its operations present convincing evidence of satisfactory performance upon this score. Before we embark upon such an examination of the operations of the company, it is requisite that we establish a frame of reference, by presenting a picture of the world banana market as a whole.

Production and Consumption

The Food and Agriculture Organization of the United Nations estimates world banana production in 1955 at 11.6 million metric tons or about 25.7 billion pounds. It apportions about 46 percent of this to South America, 23 percent to Central America, 23 percent to Asia, 6.5 percent to Africa, and 1.5 percent to Oceania. As such, the banana crop is the fourth largest of the world's reported fruit crops, exceeded in tonnage production only by grapes (88.0 billion pounds), by citrus fruit (39.2 billion pounds), and by apples (29.1 billion pounds).

Table. World Banana Production

Producing area	Million Pounds	per cent of total
Central America	5,950	23.2
South America	11,700	45.6
Asia	5,950	23.2
Africa	1,650	6.4
Oceania	400	1.6
World Total	26,650	100.0

If the portion of the grape crop produced for making wine (71.4 billion pounds) rather than for consumption as fruit is deducted, and that of apples produced for cider (8.8 billion pounds), bananas displace both grapes and apples and are second only to citrus among the world's fruit crops consumed directly as food. Excluding grapes, citrus, and apples, the tonnage of bananas exceeds by a considerable margin the combined weight of the remaining important fruit crops of the world—pears, pineapples, dates, and figs (17.2 billion pounds, combined, for 1955).

Table given above covers a world population for 1955 estimated at 2,490 million persons. On this basis, the average consumption of bananas by each man, woman, and child was 10.3 pounds in 1955 on a stem-weight basis. Only the subcategory of citrus fruits, comprising oranges, tan gerines, and clementines shows a larger per capita consumption, 12.6 pounds; and only apples, 8.1 pounds, and grapes, 6.6 pounds, were serious rivals of the banana among the world's consumption fruits.

It is, of course, obvious that such blanket averages have little relationship to the actual pattern of fruit consumption by the world's populace. Banana production is restricted to the tropics and more particularly to humid tropic areas. For successful growth, most varieties require a temperature range between 55° and 105° F., and almost all suffer severe damage when temperatures drop below 50° to 52° F.

Since the banana cannot be grown as a seasonal crop—its cycle of development ranges from 12 to 15 months from planting to harvest—its range of cultivation is restricted to the zones in which year-round temperatures are within these extremes. It is further restricted to areas that can provide its exceptionally high moisture requirements (four to five inches monthly or 60 inches annually as a minimum) through heavy and evenly spaced rainfall or generous water supplies for irrigation.

The banana thrives best in alluvial, well-drained soils, though it will tolerate clay soils of friable consistency. Since its stalk is merely a tightly rolled cylinder of leaf sheaves, it is particularly vulnerable to blowdowns or uprooting by floods when heavy with fruit. Even relatively mild winds can seriously injure the fruit by shredding the protecting leaf structure upon which healthy fruit production depends. The areas that can grow bananas successfully upon a commercial basis must therefore provide the necessary environmental factors of temperature, moisture, soil characteristics, and freedom from damaging wind and flood recurrence.

In general, the areas that can provide a hospitable environment for banana culture are found in low-lying, high-precipitation lands between 20° North and 24° South latitude, with the extremes set by 24° North and 30° South latitude. The range of actual banana production is indicated by the areas listed in Table given above.

It might be assumed that the areas of important production would be the areas of major consumption. Yet, as indicated in Table given below, some of the broad areas of banana production are heavy banana consumers as well, while others are not. On a per capita basis, the highest level of consumption in 1955 was in South America, 80 pounds per capita; while Central America (including Mexico and islands of the West Atlantic) consumed about 52 pounds per capita; Asia less than 7 pounds; Africa about 1.5 pounds; and Oceania, 27 pounds. In the United States and Canada combined consumption of bananas was 20 pounds per capita, and for Europe it was 9 pounds—an average of almost 13.5 pounds per capita for the two combined.

Thus, up to the present time, it is fair to say that bananas have assumed as important a place in the average diet in the two major industrialized areas of the temperate zone as for the vast majority of the people living in what we regard as tropical areas. In India, for example, the apparent annual per capita consumption of bananas is only about 11 pounds. Since there is so little inter-country trade in bananas within the tropic zones, it follows that there is an extraordinarily high annual per capita consumption in some of the major producing countries, while in other tropical countries the consumption rate of bananas is very low.

For 1955, on the basis of *reported* production minus export figures, Brazilians consumed something like 150 pounds per capita; Costa Ricans, 350 pounds; Panamanians, 380 pounds; and Ecuadorians a fantastic 500 to 600 pounds per capita. Even allowing for apparent inflation in some of the reported production figures and for the known fact that there is considerable feeding of bananas to livestock in some of these countries, the human consumption of bananas where they are grown in profusion clearly reaches heroic proportions.

International Trade in Bananas

Of the estimated 25.7 billion pounds of bananas produced for commercial sale in 1955, something over 25 percent was exported in international trade. The breakdown by areas of export and import, is adapted from figures compiled by the U. S. Department of Agriculture. We have adjusted the North American import weights slightly upward upon the basis of evidence that the official estimates of the Department of Agriculture do not give sufficient allowance to the increases in average stem weight in recent years. This has the effect of raising the world total of 1955 imports from the Department's estimate of about 6.6 billion to 6.7 billion pounds.

It is probable that, on the export side, there is a comparable degree of understatement of weights for fruit going to North America but we have not attempted to make this adjustment in the data presented. The discrepancy is not large enough to alter substantially the proportionate shipments as reported by broad area. Reported import and export figures never are in exact balance because of inevitable inaccuracies in accounting and because, due to time consumed in ocean transport, there are always some shipments credited as exports at the end of one year that only show up as imports when delivered at the beginning of the following year.

On the export side, the data show that almost 80 percent of all reported banana exports in 1955 were shipped from Middle America, including Mexico and the islands of the West Atlantic, and from the three South American banana exporters—Ecuador, Colombia, and Brazil. Africa accounted for about 20 percent; Asia (entirely from Taiwan) for a bit more than 1 percent; and Oceania for less than 1 percent. It is clear, then, that Latin America is the dominant bananaexporting area of the world, and that the West Coast of Africa (in which the tabulation includes the Canary Islands) is the only other major area that plays a significant exporting role, and that a relatively minor one, in the overall picture.

On the import side, North America (United States and Canada) and Western Europe, neither of which produces any banana on a commercial scale, are the outlets for over 90 percent of world banana trade. Argentina, Chile, and Uruguay are the principal Latin American importers, with Brazil and Ecuador as sources of supply. The breakdown of world banana imports for 1955 is shown below.

Of the North American imports, about 92 percent went to the United States and a little more than 8 percent to Canada, most of the latter transshipped in bond from U. S. ports. Of the three South American importing countries, Argentina takes more than 80 percent of the total. African imports are divided among the Union of South Africa, Algeria, French Morocco, Southern Rhodesia, and Tunisia in the order of their importance as banana importers. Japan is the only listed Asian importer, with about 54 million pounds imported in 1955; and New Zealand the sole importer in Oceania at a slightly lower level than Japan.

Table. World Imports of Bananas, 1955.

Importing area	Billion lbs. imported	per cent of world imports
United States and Canada	3.60	53.8
Europe	2.47	36.9
South America	.43	6.4
Africa, Asia, Oceania	.20	2.9
Total	6.70	100.0

It now can be seen that this remarkable tropical fruit, originating in the Far East and for thousands of years rarely even listed in the learned chronicles of the West, has undergone a dramatic metamorphosis—largely within the last century. Almost in the manner of historic glacial drifts, it has crept from its original domicile in the East into western zones, until today of all bananas grown, the roots of 70 percent thrust down in the soil of the Western Hemisphere.

Another 5 percent grows along the African shores that rim the Atlantic. Considered in marketing terms, within the short span of the past 60 years this exotic fruit has burst the containing bounds of its central tropic environment, and has moved northward, and to a lesser extent to the south, to find a place in the fruit bowls of temperate zone tables from the Yukon to California, from Scandinavia to Austria and Italy, and from Chile to Uruguay.

Agriculturally, bananas are chained to the tropics. Commercially, about one-fourth of all that are produced find a market outside the temperature zones in which they thrive. And nine-tenths of this fourth find their way to the industrial heart of the western world-to North America and Western Europe, whose inhabitants a century ago scarcely knew what a banana was, except for the scholarly few who read the works of Pliny the Elder or of the Swedish botanist Linnaeus. An important part of our story is concerned with how this remarkable change came to be. It is the more remarkable in that it is a phenomenon that has occurred on a comparable scale with respect to no other delicately perishable, definitely tropical fruit.

The general picture of major world fruit exports, as the average exports in metric tons, 1951-53, is adapted from a study by Erik Mortenson, entitled "Trends in Production and Consumption of Fruit and Vegetables," in the September 1955 issue of the FAO publication, *Agricultural Economics and Statistics*. The table, which was evidently compiled on the basis of official FAO trade statistics, would appear to give a reasonable basis for comparing the relative magnitudes of the reported trade in the several fruits that bulk large in international commerce.

On this accounting, the trade in bananas for 1951-53 accounted for over 36 percent of all world commerce in fruit and for more than 40 percent of world trade in fresh fruits. Oranges and tangerines make up the only other fruit category that is of even proximate importance as a world trade item upon a weight basis, and the amount of bananas traded is about 25 percent larger. Citrus fruit, however, hardly classifies as a tropical fruit. Again using FAO figures, the 1954 record shows that almost two-thirds of commercially produced oranges and tangerines were grown in temperate zone areas, more than 50 percent in the United States and Western Europe alone.

Temperate zone countries furnish 70 percent of all exports. Over 85 percent of commercial grapefruit culture is restricted to the United States, and three-quarters of all commercially grown lemons and limes are produced in the United States, Europe, Japan, Argentina, and Uruguay.

Aside from the fact that citrus fruit is far easier to ship than bananas, the striking fact is that all citrus shipments from *tropical areas* amount to only about one-quarter, by weight, of banana shipments, all of which originate in tropical areas.

Pineapples qualify as a tropical fruit crop, but the FOA figures show that only about 10 percent of total production is exported in world trade, and the great bulk of this was shipped in canned rather than fresh fruit form. Canned pineapples are included in the fresh fruit category upon the basis of the equivalent fresh fruit content. Dates are preponderantly a tropical area crop, and about one-fifth of reported production is exported internationally. But the big export is in dry rather than fresh fruit form and, in 1954, world trade in dates amounted to less than one-ninth that of bananas on a tonnage basis. Figs are the other entry on the list that in most people's minds would fall into the classification of tropical fruit. Actually, FAO figures assign considerably over half of commercial fig culture to Western Europe and the United States, and total fig exports, again with shipments preponderantly in dried form, account for less than I percent of world fruit trade. How did it come about that the banana (which is technically a herbaceous vegetable) should have become the one tropical "fruit" that has become a major item in the diet of North Americans and Europeans? Why the banana among the literally scores of fruits that are native to the tropics? Why not the mangosteen of which poet-gourmets have written in ecstatic terms, or the luscious mango, the peptic papaya, or the delicately flavoured naranjilla? It is certainly not because of special qualities that make the banana easy to grow or to transport.

We already have pointed out that the banana is singularly demanding with respect to the temperature, soil, precipitation, drainage, and wind conditions of its environment. The areas that can offer an even proximate optimum of all of these factors are relatively limited. As we shall see, bananas are subject to blights of devastating intensity for which it is tantalizingly difficult and formidably expensive to find adequate controls. This is virtually a seedless plant, with new cultivations established by planting large pseudobulbs or rhizomes, and there is no wood stalk to permit grafting.

Thus, there has been no success to date in developing, through the genetic approach, man-made strains that would combine the disease-resistant characteristics of certain varieties with the better qualities for handling and ripening inherent in others that are vulnerable to disease. In addition to disease blights, insect pests that destroy plant and fruit abound wherever bananas are grown.

The banana is a bulky fruit, shipped attached to a stem that accounts for about 7 to 8 percent of worthless weight, from which the fruit fingers protrude in a fashion that invites crushing and bruising under anything less than the tenderest handling. It must be cut in a green state, before full maturity—how long before depends upon the length of haulage to ultimate market—and the permissible time span between cutting and eating by the consumer is limited to not more than 21 to 25 days.

Yet, bananas regularly travel from 2,000 to 6,000 miles by water and up to 1,500 additional miles by rail and truck to reach their ultimate markets. Any accumulation of dust or gravel before or during transit, any roughness or cramping in stowing or carriage, even fingering by customers on the retailer's shelves shows up in marring discolouration of the delicate, golden skin of the ripened fruit and depreciates its marketability. Strict temperature controls have to be maintained from the time of loading aboard ship to delivery to retailer, and the latter is under strong compulsion to sell his fruit to customers within 24 hours after receipt.

Every stage, from plantation to retail sale, requires meticulous planning and coordination, upon a time schedule far more precise than pertains to any other major commodity of world trade. Supply in every market must be geared to a demand that, in turn, is affected by the availabilities of competitive fruits.

The cutting by maturity grade, the transport to shipping port, the stowing, the ocean carriage, the unloading, the sorting (by size, condition, and degree of ripeness), the shipment by truck or rail to distribution centres, the operations in the jobbers' ripening rooms, the trucking to retailers, and the sale to customers—every one of these stages is a tailored process with minimal tolerances for departure from a schedule that has to be calculated from its beginning. Even small divergences from the rigid timetable and handling requirements result in losses; whereas major departures would spell a total loss for the shipments involved.

If Captain Lorenzo Baker, Minor Keith, Thomas Hart, Andrew Preston, and even the imaginative Sam Zemurray had been able to visualize the complications that were to beset the production and marketing of bananas as a large-scale item in world trade, it is improbable that they would have had the fortitude to launch and expand the United Fruit Company. But if they had not made the start, it is probable that this thoroughly implausible commodity—implausible in the sense of its inherent lack of adaptability to the hazards of world marketing—would have bulked no larger in temperate zone food economies than do pineapples, or figs, or dates today.

Happily, for the North American and European consumer, and for the Latin American banana-producing economies as well, the vision of these pioneers was too limited to foresee the involvements or, at least, too sanguine to sense the formidable impediments that were to arise. So they built the foundations of today's giant trade, largely unaware of the whirling windmill blades before which even a Don Quixote might have quailed.

The Valuation of Market

The money value of bananas that enter into world trade, to the best of our knowledge, has had no systematic study. Values are commonly attributed to exports on an f.o.b. basis, and to imports on a c.i.f. basis. Such reporting is generally made by applying to the number of stems involved in either case a formula price that usually grossly understates the prices at which sales are actually made. Accordingly, the reported export valuation for a given country as shown in its annual trade account figures often represents 50 percent, or even less, of what it actually receives upon its adjusted accounts.

The International Monetary Fund, in its work of keeping track of international balance of payment flows, has been forced to make radical adjustments in reported banana trade figures to avoid untenably large distortions in the accounts of the major banana exporting countries. But the Fund only makes its adjustments for those countries in which banana exports account for a major portion of foreign exchange revenues.

The authors of this study, in attempting to put valuation figures on the world banana trade as a whole, have started with the North American market about which a good deal is known and which represented 54 percent of the world's total banana imports in 1955.

We have less complete data on the European marketing structure, which represented 37 percent of 1955 imports by weight, but there is sufficient evidence to indicate that an extrapolation of the American cost structure to that segment will not give any upward bias to the whole. The costs per pound of European bananas are consistently higher from original purchase in the producing areas on through the transportation and distribution chain. The remaining segment of world trade in bananas is too small to importantly distort the whole structure.

On a stem basis, the United Fruit Company accounted for 59 percent of North American imports, for 10 percent of European and for 28 percent of the world total. But on a weight basis—and bananas are sold by weight in markets of distribution—there are indications that the percentages of United Fruit's shipments are somewhat higher in each case, since the stem weight of the company's shipments generally runs above market averages.

Our specific data gleaned from the United Fruit Company's records furnished a far more precise base for the portion of the field that it covered than anything available in published studies. To this we added everything that could be furnished by public officials and private banana operators in the six countries in which we made detailed field surveys. As has been noted, these six, between them, produced about 60 percent of the bananas shipped in international trade in 1955. In sum, while our global estimates have been pieced together from these sources, and then blown up to represent the whole market as revealed by world trade and production statistics, we believe that our findings reasonably represent the general magnitudes of a field in which there has been no previous basis for any general appraisal.

It is our finding that, in 1955, North American consumers spent about $527 million for bananas—3,063 million pounds consumption weight; 3,604 million pounds stem weight—at an average retail price of 17.2¢ per pound. We know that Europeans paid a somewhat higher price per pound than Americans, and that other importing areas paid somewhat less. Taking the North American price as average, and knowing that this area accounted for 54 percent of all imports, we can estimate the world retail expenditure for *imported* bananas at approximately $976 million. We have seen earlier that about three-quarters of the reported world commercial production of bananas is consumed in the countries where they are produced. We have no overall reporting of what consumers paid for bananas in the producing countries, but we have some data on this for our six-country sample. In these countries, bananas sold on the local market for from one-fifth to one tenth of what they brought when sold as export stems.

It is certainly conservative to estimate that the totality of fruit grown for local markets throughout the world, although almost three times as great in quantity, has less than one-half and perhaps not more than a third the value of that committed to export markets. It is probably safe to estimate the total retail value of the 1955 world's commercial banana crop at between $1.3 billion and $1.5 billion.

We are able to make a reasonably proximate estimate of the value in 1955 of world banana exports f.o.b. vessels at ports of embarkation. On the basis of our North American consumer dollar analysis that follows, it will be seen that about 27¢ out of each such dollar spent on bananas represents the return actually realized by producing countries. Hence, of the estimated $976 million sales at retail, $263 million can be assigned as the share of the producing countries from their banana exports.

Mango processing technologies

Mangoes are processed at two stages of maturity. Green fruit is used to make chutney, pickles, curries and dehydrated products. The green fruit should be freshly picked from the tree. Fruit that is bruised, damaged, or that has prematurely fallen to the ground should not be used. Ripe mangoes are processed as canned and frozen slices, purée, juices, nectar and various dried products. Mangoes are processed into many other products for home use and by cottage industry.

The mango processing presents many problems as far as industrialization and market expansion is concerned. The trees are alternate bearing and the fruit has a short storage life; these factors make it difficult to process the crop in a continuous and regular way. The large number of varieties with their various attributes and deficiencies affects the quality and uniformity of processed products.

The lack of simple, reliable methods for determining the stage of maturity of varieties for processing also affects the quality of the finished products. Many of the processed products require peeled or peeled and sliced fruit. The lack of mechanised equipment for the peeling of ripe mangoes is a serious bottleneck for increasing the production of these products.

A significant problem in developing mechanised equipment is the large number of varieties available and their different sizes and shapes. The cost of processed mango products is also too expensive for the general population in the areas where most mangoes are grown.

There is, however, a considerable export potential to developed countries but in these countries the processed mango products must compete with established processed fruits of high quality and relatively low cost.

Green Mango Processing

Pickles

The optimum stage of maturity should be determined for each variety used to make pickles. There are two classifications of pickles - salt pickles and oil pickles. They are processed from whole and sliced fruit with and without stones. Salt is used in most pickles.

The many kinds of pickles vary mainly in the proportions and kinds of spices used in their preparation. One basic recipe for the study of the preparation and storage of pickles in oil is as follows:

Mango pieces	250 g	Tumeric powder	2 to 4 g
Salt	60 g	Fenugreek seeds	2 to 4 g
Mustard powder	30 g	Bengal gram seeds	2 to 4 g
Chili powder	20 g	Gingelly oil	20 to 30 g

The ingredients are mixed together and filled into wide-mouthed bottles of 0.5 kg capacity. Three days later the contents are thoroughly mixed and refilled into the bottles. Extra oil is added to form a 1-2 cm layer over the pickles.

Chutney

The product is prepared from peeled, sliced or grated unripe or semi-ripe fruit by cooking the shredded fruit with salt over medium heat for 5 to 7 minutes, mixed and then sugar, spices and vinegar are added. Cook over moderate heat until the product resembles a thick purée, add remaining ingredients and simmer another 5 min. Cool and preserve in sterilised jars.

Spices usually include cumin seeds, ground cloves, cinnamon, chili powder, ginger and nutmeg. Other ingredients such as dried fruits, onions, garlic and nuts may be added.

Drying/dehydration. immature fruit is peeled and sliced for sun-drying. The dried mango slices can be powdered to make a product called amchoo. The use of blanching, sulphuring and mechanical dehydration gives a product with better colour, nutrition, storability and fewer microbiological problems.

Ripe Mango Processing

Puree

Mangoes are processed into purée for re-manufacturing into products such as nectar, juice, squash, jam, jelly and dehydrated products. The purée can be preserved by chemical means, or frozen, or canned and stored in barrels. This allows a supply of raw materials during the remainder of the year when fresh mangoes are not available. It also provides a more economical means of storage compared with the cost of storing the finished products, except for those which are dehydrated, and provides for more orderly processing during peak availability of fresh mangoes.

Mangoes can be processed into purée from whole or peeled fruit. Because of the time and cost of peeling, this step is best avoided but with some varieties it may be necessary to avoid off-flavours which may be present in the skin. The most common way of removing the skin is hand-peeling with knives but this is time-consuming and expensive. Steam and lyepeeling have been accomplished for some varieties.

Several methods have been devised to remove the pulp from the fresh ripe mangoes without hand-peeling. A simplified method is as follows: the whole mangoes were exposed to atmospheric steam for 2 to 2 1/2 min in a loosely covered chamber, then transferred to a stainless steel tank.

The steam-softened skins allowed the fruit to be pulped by a power stirrer fitted with a saw-toothed propeller blade mounted 12.7 to 15.2 cm below a regular propeller blade. The pulp is removed from the seeds by a continuous centrifuge designed for use in passion fruit extraction. The pulp material is then passed through a paddle pulper fitted with a 0.084 cm screen to remove fibre and small pieces of pulp.

Mango purée can be frozen, canned or stored in barrels for later processing. In all these cases, heating is necessary to preserve the quality of the mango purée. In one process, purée is pumped through a plate heat exchanger and heated to 90°C for 1 min and cooled to 35° C before being filled into 30 lb tins with polyethylene liners and frozen at -23.50 C.

In an other process, pulp is acidified to pH 3.5, pasteurized at 90°C, and hot-filled into 6 kg high-density bulk polyethylene containers that have been previously sterilised with boiling water. The containers are then sealed and cooled in water. This makes it possible to avoid the high cost of cans.

Wooden barrels may be used to store mango pulp in the manufacture of jams and squashes. The pulp is acidified with 0.5 to 1.0per cent citric acid, heated to boiling, cooled, and SO2 is added at a level of 1000 to 1500 ppm in the pulp. The pulp is then filled into barrels for future use.

Slices

Mango slices can be preserved by canning or freezing, and recent studies have shown the feasibility of pasteurized-refrigerated and dehydro-canned slices. The quality of the processed product in all of these procedures will be dependent upon selection of a suitable variety along with good processing procedures. Thermal process canning of mango slices in syrup is the most widely used preservation method.

Beverages

The commercial beverages are juice, nectar and squash. Mango nectar and juice contain mango purée, sugar, water and citric acid in various proportions depending on local taste, government standards of identity, pH control, and fruit composition of the variety used.

Mango squash in addition to the above may contain SO2 or sodium benzoate as a preservative. Other food grade additives such as ascorbic acid, food colouring, or thickeners may be used in mango beverages. A short description of finished products found in literature is as follows:

- Mango juice: prepared by mixing equal quantities of pulp (purée) and water together and adjusting the total soluble solids (TSS) and acidity to taste (12 to 15per cent TSS and 0.4 to 0.5per cent acidity as citric acid);
- Mango nectar containing 25per cent purée can be prepared using the following procedure.

Table. Brix of Puree.

Nectar components	15°	17°	20°
Purée	100	100	100
Sugar	45	43	40
Water	255	257	260

Commercial processing conditions may require the use of a preservative.

The pH is adjusted to approximately 3.5 by adding citric acid as a 50per cent solution.

The time of heat processing will vary with filling temperature, can size and viscosity of the juice or nectar.

Mango squash may be prepared according to flow-sheet described below; the finished product may contain 25per cent juice, 45per cent TSS and 1.2 to 1.5per cent acidity and may be preserved with sulphur dioxide (350ppm) or sodium benzoate (1000 ppm) in glass bottles. Mango squash simplified flow-sheet.

Ingredients

Mango pulp	900	900
Sugar	900	1100
Citric acid	18	15
Water	900	900

Mangoes are washed, stored, peeled with stainless steel knives. The pulp is prepared by using a pulper with fine sieve (0.025-in); Sugar is mixed with water and citric acid = syrup; The pulp is added to the syrup and mixed well; The mixture is strained trough cloth; The squash is heated at 85° C and bottles are filled and closed.

For additional heat treatment bottles may need to be maintained at a product temperature of 80°C for 30 minutes if the product is to be processed without preservatives. The bottles are then left to cool in water and stored at room temperature.

Two negative points must be avoided: presence of air bubbles (which is a source of quick deterioration) and separation of squash solids (giving an undesirable appearance). The means to avoid these two phenomena are described in the fruit juices section.

A type of "squash type" beverage may also be manufactured with '/a pulp + '/a water + i/a sugar and pH adjusted to 3.7 by addition of citric acid. Using different sieve sizes affects the quality and reduces air bubbles to a certain extent but homogenisation and de-aeration of purée or squash seem to be important in order to avoid separation and air bubbles. The squash quality is evaluated on the basis of the following characteristics: pH, titrable acidity, soluble solids, ascorbic acid (by 2,6 dichlorophenol indophenol method), specific gravity.

Dried/Dehydrated

Ripe mangoes are dried in the form of pieces, powders, and flakes. Drying procedures such as sun-drying, tunnel dehydration, vacuum-drying, osmotic dehydration may be used. Packaged and stored properly, dried mango products are stable and nutritious.

One described process involves as pre-treatment dipping mango slices for 18 hr (ratio 1:1) in a solution containing 40° Brix sugar, 3000 ppm SO_2, 0.2per cent ascorbic acid and 1per cent citric acid; this method is described as producing the best dehydrated product. Drying is described using an electric cabinet through flow dryer operated at 60° C. The product showed no browning after 1 year of storage.

Drum-drying of mango purée is described as an efficient, economical process for producing dried mango powder and flakes. Its major drawback is that the severity of heat preprocessing can produce undesirable cooked flavours and aromas in the dried product. The drum-dried products are also extremely hydroscopic and the use of in-package desiccant is recommended during storage.

Canning

This preservation technology is described in various technological flow-sheets in this bulletin. Mango bar or "fruit leather" is presented in various flow-sheets.

Mango Malformation

It is the most important malady of mango and was first reported by Burn (1910). It causes the losses from 50-80 per cent. It is also reported to be due to other disorders like physiological and acarological. Indications of its virus nature has also been postulated. The maximum incidence of the disease has been recorded in the north, north-west and north-east parts of the country. The incidence is, however, less in western and southern India. There are two types of symptoms, namely, floral malformation and witches broom or bunchy top or vegetative malformation with a proliferation of infected tissue. The flowering panicles turn into a compact mass of flowers.

This compact mass is very hard and not soft like normal panicle. Individual flower is greatly enlarged and has a large disc. The percentage of bisexual flowers in malformed panicles is very low. In bunchy top, compact leaves are formed in a bunch at the apex of shoot or in the leaf axil. A similar bunch consisting of small rudiments crowded together on short shootlets is seen in vegetative malformation in which the growth of shootlet is arrested. Vegetative malformation is more pronounced in young seedling and seedling trees. The malformed heads dry up in black masses and persist on the trees for a long time. The malformed inflorescences contain more of endogenous cytokinins than healthy ones (Nicholson and van Staden 1988).

The variety Zebda has been reported to be resistant in Egypt (El Ghandour *et al.* 1979). Considerable incidence reduction has been reported by spraying 100-200 ppm NAA during October.

Use of disease free planting material and prophylactic spray of insecticides and fungicides can keep the orchards healthy. In areas with less than 5-10 per cent infection, pruning of diseased plants should be made compulsory; pruning of diseased parts along the basal 15-20 cm apparently healthy portions. This is followed by the spraying of Bavistin (0.1per cent) or Captaf (0.2per cent). Partial control of malformation has been achieved by spraying mangiferin-Zn2 and mangiferin-Cu2+chelates.

Stem End Rot (Botryodiplodia theobromae Pat.)

It is an important disease of ripe mango. This fungus also causes die-back symptoms in orchards. It is known to occur in India, Myanmar, Sri Lanka, the Philippines, Mauritius and the U.S.A. In India, it is recorded from all the mango growing areas. The disease is characterized as the dark epicarp around the base of the pedicel in the initial stage. In the next few hours, the affected area enlarges to form a circular, black patch which in humid atmosphere extends rapidly and turns the whole fruit completely black within two or three days. The pulp of the diseased fruits become brown and somewhat softer. The diseased fruits showed increased phosphorus and reduced ascorbic acid contents. Pectolytic enzyme activity was also lowered in such fruits. The disease is carried from the dead twigs and bark of the trees harbouring the pathogen and with the onset of rains the orchards get contaminated.

Dipping of mangoes in 6 per cent borax solution at 43°C for 3 minutes reduces the incidence. Fruit dipping in Aureofungin and wrapping was also found effective.

Pathak (1980) suggested to harvest mangoes on clear dry day. Hot water treatment at 53°C for 10 minutes and at 51 to 55°C for 15 minutes have been recommended elsewhere. Care should be taken to prevent snapping-off of the pedicels. Injury should be avoided to fruits at all stages of handling. Spalding (1982) in his experiments found thiobendazole and thiophenate methyl to be the most effective ones. Recently, Rawal and Ullasa (1988) have reported that Bavistin (0.1per cent) or Topsin-M (0.1per cent) or chlorothalonil (0.2per cent) spraying in the field before harvesting give very effective control.

Die-back (Botryodiplodia thiobromae Pat)

It is a descructive disease of mango and is known to occur in India and other mango growing countries. The disease is manifested by discolouration and darkening of the bark some distance from the tip of the twigs. The dark area advances and young green leaves start withering first at the base and then extending outwards along the vein. The affected leaves turn brown and the margins roll upward. At this stage, the twig or branch dies, leaves shrivel and fall, and this may be accompanied by exudation of gum. The infected twigs show internal discolouration. Brown streaks of vascular tissues is seen on splitting the twigs lengthwise along the long axis. In early stages, epidermal and sub-epidermal cells of twigs appear slightly shrivelled. The areas of cambium and phloem show brown discolouration and yellow gum like exudate flows out of the cells. The disease is known to be enhanced by a beetle, Xyleborus affinis (Batista 1947). Relative humidity above 80per cent and temperature of 25 to 31.5°C, and rains enhance disease development. The die-back is also caused due to Colletotrichum gloeosporioides. Both these pathogens also attack fruits which get spoiled during ripening.

For effective control of the disease, pruning and destruction of infected twigs is the foremost practice. Spraying the trees periodically with copper oxychloride sulphate is recommended by Alvarez and Lopez (1971). Pasting of trees with a mixture of oil and 5per cent phenol was found effective. Sprays of carbendazim (0.1per cent) or methyl thiophenate (0.1per cent) or chlorothalonil (0.2per cent) at fortnightly interval during rainy season is important. Sometimes short hole borers also predispose the trees to infection and hence proper insecticides are also to be sprayed. Healthy twigs should be selected for grafting of seedling during propagation.

Leaf Blight (Macrophomina mangiferae Hingorani and Sharma)

It was first reported from Brazil to be caused by Macrophomina sp. by Lacca (1922). In India, it was reported by Patel *et al.* (1949). Later, it has been reported by various workers from Delhi, U.P. and Gujarat. The other pathogens causing the disease are grey blight (Pestalotiopsis mangiferae), phoma blight (Phoma glomerate), and twig blight and fruit rot (Phoma sp.). The disease usually appears as yellowish pin head like spots on leaves and twigs of the affected plants. These gradually enlarge discolouring the surrounding tissues which first become brown and then dark brown with slightly raised and brown purplish margins and later ash coloured due to the appearance of pycnidia.

Spots are round to start with but later become oval or irregular in size depending upon environmental conditions such as humidity and temperature. Infection is mainly observed on leaves and rarely on stems. On stem, the lesions are elliptic which later girdle the stem at the point of infection. On fruits, water soaked circular lesion are produced which enlarge rapidly and cause rotting.

The pathogen can survive for more than a year on the leaves of mango. The varieties, namely, Khandeshi, Borasio and Asal Damido are reported to be resistant. Field sanitation is by collecting and destruction of diseased parts. Spraying of Burgandy mixture, Perenox, lime sulphur and dithane have been recommended by Hingorani *et al.* (1960).

Red Rust (Cephaleuros mycoides Karst)

This or red spot is a common algal disease on the mango in the tarai and in the other humid regions of India. Its occurrence has been reported from Bihar, Karnataka and Uttar Pradesh.

The alga attacks the foliage, bark and twigs of the host plant and is both parasitic and epiphytic in nature. In serious infections, the bark becomes thickened, the twigs get enlarged and remain stunted and the foliage becomes sparse and finally dries up. Initially the spots are greenish-grey in colour and velvety in texture, but later on the surface bears reddish-brown appearance. Algal spot is circular to irregular in shape, slightly elevated and with usually 2 mm in diameter, though in some cases it may be as much as 1 cm. Bordeaux mixture (6:6:100), Supramar, Fytolan, Blitox-50 or lime sulphur are recommended.

Black Spot (Xanthomonas compestris pv mangiferae indicae)

It is one of the serious diseases and is reported from many countries. In India, it is reported from Maharashtra, Delhi, Uttar Pradesh, Tamil Nadu and Karnataka. It affects leaves, petioles, fruits and tender stems. Numerous small angular water soaked lesions appear in groups towards the tip of the leaf blade. These are first light yellowish but later turn dark brown to black and get surrounded by a distinct halo. Several lesions are often found to coalesce forming large necrotic spots. On young fruits, water-soaked lesions develop which also turn dark brown to black. Infected fruits may show skin cracking and badly affected ones dry prematurely. Infection by wounding produces diseased spots on tender stems, petioles and stalk bearing fruits.

These spots turn into deep longitudinal scars accompanied by appreciable amount of bacterial gum-like ooze. The incubation period on leaves varies from 2-14 days. In India, the disease remains dormant during November to March due to low temperature (11.8-22°C) and the leaf infection is considerably reduced by the fall of the infected leaves. Kent mangoes showed close relationship of rainy season with incidence of bacterial leaf spot. One day's rainfall did not affect the disease appearance, whereas four days of rainfall led to the 36.9per cent disease incidence.

Twig canker persists and initiates fruit infection. The disease spread is rapid during the rains and becomes severe in July-August. Five application of Bordeaux mixture (4:4:50) with spreader was recommended by Wager (1937), whereas four sprays of 6:6:50 Bordeaux mixture was found to be effective by Marloth (1947). Sprays with fungicides such as copper containing materials and agrimycin have also been advocated. Shekhawat and Patel (1975) recommended orchard sanitation by way of removal of infected materials and seedling treatment as preventive measures. Agrimycin-100 proved best in arresting the disease development. Plantomycin (200 ppm) followed by agrimycin + bavistin (1000 ppm) was reported to be the best against this disease. Streptomycin sulphte followed by aureofungin have been recommended by Prakash and Raooff (1985) to control bacterial disease of mango.

Black Mould Rot (Aspergillus niger Van Teigh)

The affected fruits show yellowing of base and development of irregular, hazy, greyish spots and coalesce into dark brown or black lesions. The mesocarp of the rotted areas becomes depressed and soft. The diseased fruits show rapid decrease in ascorbic acid content. The pathogen infects injured fruits. Totapuri variety is highly susceptible. A fruit dip treatment with benlate at 1500 ppm can control the rot.

Soft Rot (Rhizopus arrhizus Fisc.)

This pathogen alone is reported to be responsible for the decay of 6.3 per cent fruits in Delhi markets. The pathogen incites a typical soft rot. Development of disease is slow and less intensive at low temperature (7-8°C). The disease develops rapidly between 20 and 40°C. Thakur and Chenulu (1970) found that 2-aminothiozole with 2-aminopyridine each at 7 per cent concentration protected mango fruits for 20 days. Some of the other post-harvest diseases and pathogens are Alternaria rot (Alternaria tenuissima (Neer. Ex. fr.) Wiltshire), Sclerotium rot (Sclerotium rolfsii Sacc.), dry rot (Boothiella tetraspora Lodhi and Mirza), brown rot (Pestalotiopsis mangiferae P. Henn. and P. glandiola (Cast) Stey), fruit rot (Phomopsis mangiferae Ahmed) and black spot, Actinodichium jenkinsii.

Black Tip

This disease is peculiar to India and does not occur in any other country. It has been reported to occur in regions of West Bengal, Uttar

Pradesh, Bihar and Punjab in orchards located in close proximity of a brick kiln. South India is free from this disease. The symptoms manifest as small etiolated area at the distal end of the fruit, which gradually spreads, turns nearly black and covers the tip completely. Before the etiolation is complete, isolated greyish spots appear which become dark brown, enlarge and coalesce into a continuous necrotic area. The brown discolouration spreads to the neighbouring parenchyma and the deposits also appear in the ducts. The browning and deposits gradually spread throughout the mesocarp and the affected cells distintegrate and coalesce into a dead tissue. In severe cases, the necrosis extends to the endocarp. The practical control measure lies in keeping the brick kilns away from the mango orchards. The use of telescoping chimney 12-15 m high has also been recommended. Spraying of borax (0.6per cent) at 10-14 day intervals has been recommended.

Soft Nose

The problem was first observed as physiological breakdown in cvs. Kent and Haden by orchardists in 1950 in Florida. The soft nose is characterized by hallowing of the green skin in the areas between the apex and the stigmal point. A lack of firmness in diseased fruits can also be felt with experience. Upon cutting the flesh on the ventral side towards the apex of some soft nose fruits appear to be over ripe, while on shoulder and dorsal side it is unripe. The over ripe flesh surrounds a mass of yellowish to brown tissue which is firmer than the surrounding affected tissue and is bitter in taste.

Young (1957) concluded that the disorder is not due to the microorganisms. However, Young *et al.* (1963) reported the loss of only few percent in fruits harvested from trees grown on calcareous rock soil, but on acidic soil, sandy soil it is common for 15-20 per cent incidence. According to Young *et al.* (1963), soft nose disorder increased with N fertilization but was reduced to greater extent by increasing Ca level of the soil. Nitrogen fertilizer in sandy acidic soil increased the incidence whereas in calcareous soil there was no effect. To overcome this disorder, the high N level of soil should not be allowed in the sandy or acidic soil and proper Ca level should also be maintained in soil. Early picking is also suggested. The other disorders are little leaf caused by zinc deficiency, tip burn due to moisture stress and salt accumulation in soil and leaf scorch due to accumulation of chloride ion in leaf.

Phanerogamic Parasites (Dendrophthoe (Loranthus) viscum)

The mango trees are severely affected by this parasite particularly in neglected orchards. They are commonly present on the trunk or branches of the trees. The foliage of the infected tree is sparse, reduced in size and its bearing capacity and quality of fruits is considerably reduced. The fruits of this parasite are baccate with viscous mesocarp and attract the birds by its attractive colour. The seeds of this parasite are spread through bird and animal droppings. Such seeds, germinate and produce haustoria. Sometime tumour like swellings are observed on the affected tissue. There are many other species of parasites which also attack mango trees. All epiphytic plants can be removed easily in early stages before their aerial roots penetrate the ground and surround the host trunk.

Some more diseases reported by various workers attacking the mango crop in the field are leaf spots due to Rhizoctonia bataticola (Taub) Butt., Cercospora mangiferae indica Munjal Lal and Chona, Alternaria alternata Fr. and Keissler, Phoma sorghicola (Sacc.) Boerema. Dorem and Venkast; Scab (Elsinoe mangiferae Bit and Jen; black band (Rhinocladium corticolum Massee); pink disease (Pellicularia solmonicolour (Berk and Br.) Dastur); root rot and damping-off (Rhizoctonia solani Kuhn) and wilt due to Verticillium albo-atrum Reinke and Berth and Fusarium solani (Mart.) Sacc. There are several more fungi causing different diseases but the above are the most important among them.

Recent trends in fruit and vegetable processing

New Products

The number and variety of fruit and vegetable products available to the consumer has increased substantially in recent years. The fruit and vegetable industry has undoubtedly benefited from the increased recognition and emphasis on the importance of these products in a healthy diet.

Traditional processing and preservation technologies such as heating, freezing and drying together with the more recent commercial introduction of chilling continue to provide the consumer with increased choice. This has been achieved by new heating (*e.g.* UHT, microwave, ohmic) and freezing (*e.g.* cryogenic) techniques combined with new packaging materials and technologies (*e.g.* aseptic, modified atmosphere packaging).

The overall trend in new fruit and vegetable products is "added value", thus providing increased convenience to the consumer by having much greater variety of ready prepared fruit and vegetable products. These may comprise complete meals or individual components. The suitability of products and packages for microwave re-heating has been an important factor with respect to added convenience.

Fruit and vegetable product trends

Heat processed products

- Canned fruits and vegetables
  - Combination of vegetables in sauces and vegetable recipe dishes. Exotic fruits.
- Glass packed fruits and vegetables
  - "Condiverde"/"antipasti" products based on vegetables in oil.
  - High quality fruit packs.
- Retortable plastics
  - Basic vegetables or vegetable meals
  - Fruit in jelly
- Aseptic cartons
  - Ready made jelly
- Rosti meals
  - Potato based products in retort pouches
- Fruit juices
  - New combinations of juices and freshly squeezed products
- Crisps
  - Thick and crunch skin-on crisps. Kettle or pan fried chips. Lower fat crisps.

Table. Numbers of New Fruit and Vegetable Products.

	1990	1991	1992 Jan-June
Frozen vegetable products	66	95	21
Chilled vegetable products	76	81	78
Heat processed vegetables	51	60	38
Heat processed fruits	13	14	5
Fruit juices and drinks	73	83	46
Potato crisps	32	33	16

New product development in the fruit and vegetable sector is most important in meeting the continued challenge of providing the consumer with choice and high quality products.

A Fresh Look at Dried Fruit

New fruit varieties and advance in drying technologies are putting a fresh twist on dried fruit applications. Fruits that have been introduced to the drying process include cranberries, blueberries, cherries, apples, raspberries and strawberries - not to mention the traditional mainstays of raisins, dates, apricots, peaches, prunes and figs. Perceived as a "value-added" ingredient, dried fruit adds flavour, colour, texture and diversity with little alteration to an existing formula. The growing interest in ethnic cuisines in U.S.A. and the change to a more healthy way of eating, has also moved dried fruit considerably closer to the mainstream.

Found primarily in the baking industry, dried fruit is coming into its own in various food products, including entrees, side dishes and condiments. Compotes, chutneys, rice and grain dishes, stuffings, sauces, breads, muffins, cookies, deserts, cereals and snacks are all food categories encompassing dried fruit.

Since some dried fruit is sugar infused (osmotic drying), food processors can decrease the amount of sugar in formula - this is especially the case in baked products. Processors are making adjustments in moisture content of the dried fruit so that a varied range is available for different applications. An added bonus is dried fruits' shelf stability (a shelf life of at least 12 months). Dried fruit is more widely available in different forms, including whole dried, cut, diced and powders.

Citric Acid and its Use in Fruit and Vegetable Processing

Citric acid may be considered as "Nature's acidulant". It is found in the tissues of almost all plants and animals, as well as many yeasts and moulds. Commercially citric acid is manufactured under controlled fermentation conditions that produce citric acid as a metabolic intermediate from naturally-occurring yeasts, moulds and nutrients. The recovery process of citric acid is through crystallization from aqueous solutions.

Citric acid is widely used in carbonated and still beverages, to impart a fresh-fruit "tanginess". Citric acid provides uniform acidity, and its light fruity character blends well and enhances fruit juices, resulting in improved palatability. The amount of citric acid used depends on the particular desired flavour (*e.g.*, High-acid: lemonade; Medium-acid: orange, punch, cherry; Low-acid: strawberry, black cherry, grape).

Sodium citrate is often added to beverages to mellow the tart taste of high acid concentrations. It provides a cool, distinctive smooth taste and masks any bitter aftertaste of artificial sweeteners. In addition, it serves as a buffer to stabilise the pH at the desired level. The high water solubility of citric acid (181 g/100 ml) makes it an ideal additive for fountain fruit syrups and beverages concentrates as a flavour enhancer and microbial growth inhibitor (preferably at pH < 4.6).

In processed fruits and vegetables, citric acid performs the following functions:

- It reduces heat-processing requirements by lowering pH: inhibition of microbial growth is a function of pH and heat treatment. Higher heat exposure and lower pH result in greater inhibition. Thus the use of citric acid to bring pH below 4.6 can reduce the heating requirements. In canned vegetables, citric acid usage is greatest in tomatoes, onions and pimentos. For tomato packs, the National Canners Association recommends a pH of 4.1 to 4.3. In general, 0. 1per cent citric acid will reduce the pH of canned tomatoes by 0.2 pH units.

- Optimise flavour: citric acid is added to canned fruits to provide for adequate tartness. Recommended usage level is generally less then 0.15per cent.
- Supplement antioxidant potential: citric acid is used in conjunction with antioxidants such as ascorbic and erythorbic acids, to inhibit colour and flavour deterioration caused by metal-catalysed enzymatic oxidation. Recommended usage levels are generally 0.1per cent to 0.3per cent with the antioxidant at 100 to 200 ppm.
- Inactivate undesirable enzymes: oxidative browning in most fruits and vegetables is catalysed by the naturally present polyphenol oxidase. The enzymatic activity is strongly dependent on pH.
- Addition of citric acid to reduce pH below 3 will result in inactivation of this enzyme and prevention of browning reactions.

Cherry and Apricot Oils are Safe for Food Use

The oils obtained by cold pressing the kernels of the cherry (Prunus cerasus) and the apricot (Prunus armeniaca) have been declared acceptable for food use by the UK Ministry of Agriculture, Fisheries and Food's Advisory Committee on Novel Foods and Processes (ACNFP) by June 1993.

In its assessments of the safety in use of the cherry and apricot kernel oils, the ACNFP consider specifications that included data on fatty acid composition, the presence of natural antioxidants and the content of cyanide, mycotoxins and heavy metals.

The Committee says that it gave particular consideration to the possible presence in the oils of the cyanogenic glucoside amygdalin, from which cyanide is released by enzymic action when the kernels of cherry and apricot are crushed. Amygdalin was found to be absent from the cherry and apricot kernel oils. The ACNFP is satisfied that there are no food safety reasons why the use of cherry and apricot kernel should not be acceptable provided there is compliance with the specifications shown in Table given below.

Table. Specification of Purity for Cherry and Apricot Kernel Oils as Determined by UK ACNFP.

	Cherry	Apricot
Contaminants limits:		
Heavy metals (total)	0.5 mg/kg	0.5 mg/kg
Aflatoxins (total)	4.0 g/kg	4.0 g/kg

Cyanide	0.15 mg/kg	0.15 mg/kg
Pesticide residues	0.01 mg/kg	0.01 mg/kg
Tocopherols:		
Alpha/delta/gamma (mg/kg)	356-886	569-899

The oils are obtained by the mechanical mincing and cold pressing of kernels extracted from cleaned cherry or apricot stones. After filtering, the oils are stored and are to be sold in a raw, unrefined state. The cherry and apricot kernel oils are high unsaturated and are expected to be used as speciality oils for salad dressings, baking and shallow frying applications.

The Use of Fruit Juices in Confectionery Products

During the last decade, the concept of fruit juices has gained immensely on consumer popularity. The majority of new non-alcoholic and alcoholic fruit drink products were a combination of syrups, fruit juices and flavours.

The confectionery industry followed suit and new products incorporated fruit juices as part of their confectionery formulations and processes. Fruit juice concentrates of high solids are often used instead of normal or single-fold juices.

Juice concentrates are made of pure fruit juices. The process starts with pressing fruits and obtaining pure fruit juice; this is stabilised by heat treatment which inactivates enzymes and micro-organisms. The next processing step is concentration under vacuum up to 40-65° Brix or 4-7 fold. The concentrates are then blended for standardisation and stored.

These fruit juice concentrates are often further stabilised by the addition of sodium benzoate and potassium sorbate and are usually stored away from light and are refrigerated or frozen.

Depectinised fruit juices are also used to prevent foaming in confectionery processes and are essential for use in clear beverage products. Fruit juice concentrates which are depectinised, and have added preservatives are called stabilised, clarified, fruit juice concentrates.

Fruit juices are used in confectionery products in conjunction with natural and artificial flavours which provides intense flavour impact and are cost-effective for a confectionery product.

The traditional concern in using fruit juice concentrates in confectionery applications has been the effect of the natural acids on the finished product, particularly the formation of invert sugar during processing.

This is a logical concern since concentrates contain differing amounts and types of acids. For example: apple, cherry, strawberry and other berries contain primarily malic acid. Grapes mainly contain tartaric acid.

Cranberry is high in quinic acid. Citrus fruits and pineapple contain differing amounts of citric acid. The concentrates, when used, are normally buffered to a pH of 5-7 with sodium hydroxide.

In formulating products with fruit juice concentrates, the solids of the concentrate are considered as mostly reducing sugars and a reduction in corn syrup is made to compensate for equivalent amount of reducing sugar being added in the concentrate.

The exact replacement can be determined by measuring the D.E. of the concentrate to be added. In formulations when small amounts of concentrate are used (less than 1per cent), no adjustment is made since the reducing sugar contribution of the concentrate is not significant.

Fruit juice concentrates can also be used to provide a source of natural colour, in particular red colour. Grape, raspberry, cherry, strawberry and cranberry concentrates in small amounts are very effective in colouring cream centres.

The inclusion of fruit juices in confectionery products is now left up to the imagination of the manufacturer. These products must, of course, hold up to the standards of flavour integrity, and product excellence, during the shelf-life of these products.

Vegetable Specific Processing Technologies

Vegetables varieties

Vegetable processors must appreciate the substantial differences that varieties of a given vegetable will possess. In addition to variety and genetic strain differences with respect to weather, insect and disease resistance, varieties of a given vegetable will differ in size, shape, time of maturity, and resistance to physical damage.

Varietal differences then further extend into warehouse storage stability, and suitability for such processing methods as canning, freezing, pickling or drying. A variety of peas that is suitable for canning may be quite unsatisfactory for freezing and varieties of potatoes that are preferred for freezing may be less satisfactory for drying or potato chip manufacture.

This should be expected since different varieties of a given vegetable will vary somewhat in chemical composition, cellular structure and biological activity of their enzyme system.

Harvesting and pre-processing

When vegetables are maturing in the field they are changing from day to day. There is a time when the vegetable will be at peak quality from the stand-point of colour, texture and flavour. This peak quality is quick in passing and may last only a day. Harvesting and processing of several vegetables, including tomatoes, corn and peas are rigidly scheduled to capture this peak quality.

After the vegetable is harvested it may quickly pass beyond the peak quality condition. This is independent of microbiological spoilage; these main deteriorations are related to:

- Loss of sugars due to their consumption during respiration or their conversion to starch; losses are slower under refrigeration but there is still a great change in vegetable sweetness and freshness of flavour within 2 or 3 days;
- Production of heat when large stockpiles of vegetables are transported or held prior to processing.

At room temperature some vegetables will liberate heat at a rate of 127,000 kJ/ton/day; this is enough for each ton of vegetables to melt 363 kg of ice per day. Since the heat further deteriorates the vegetables and speeds micro-organisms growth, the harvested vegetables must be cooled if not processed immediately. But cooling only slows down the rate of deterioration, it does not prevent it, and vegetables differ in their resistance to cold storage. Each vegetable has its optimum cold storage temperature which may be between about 0-100 C (32-500 F).

- The continual loss of water by harvested vegetables due to transpiration, respiration and physical drying of cut surfaces results in wilting of leafy vegetables, loss of plumpness of fleshy vegetables and loss of weight of both.

Moisture loss cannot be completely and effectively prevented by hermetic packaging. This was tried with plastic bags for fresh vegetables in supermarkets but the bags became moisture fogged, and deterioration of certain vegetables was accelerated because of buildup of CO2 and decrease of oxygen in the package. It therefore is common to perforate such bags to prevent these defects as well as to minimise high humidity in the package which would encourage microbial growth.

Shippers of fresh vegetables and vegetable processors, whether they can, freeze, dehydrate, or manufacture soups or ketchup, appreciate the instability and perishability of vegetables and so do everything they can to minimise delays in processing of the fresh product. In many processing plants it is common practice to process vegetables immediately as they are received from the field.

To ensure a steady supply of top quality produce during the harvesting period the large food processors will employ trained field men; they will advise on growing practices and on spacing of plantings so that vegetables will mature and can be harvested in rhythm with the processing plant capabilities. This minimises stockpiling and need for storage.

Cooling of vegetables in the field is common practice in some areas. Liquid nitrogen-cooled trucks may next provide transportation of fresh produce to the processing plant or directly to market. Upon arrival of vegetables at the processing centre the usual operations of cleaning, grading, peeling, cutting and the like are performed using a moderate amount of equipment but a good deal of hand labour also still remains.

Reception

This covers qualitative and quantitative control of delivered vegetables. The organoleptic control and the evaluation of the sanitary state, even if they are very important steps in vegetables' characteristics assessment, cannot establish their technological value. On the other hand, labouratory controls do not precisely establish their technological properties because of the difficulty in putting into showing some deterioration when using rapid control methods. One correct method of vegetable quality appraisal is their overall evaluation based on the whole complex of data that can be obtained by combining an extensive organoleptic evaluation with simple analysis that can be performed rapidly in plant labouratory. These analysis can be:

- Refractometric extract (tomatoes, fruit, etc.);
- Specific weight (potatoes, peas, etc.);
- Consistency (measured with tenderometres, penetrometres, etc.);
- Boiling tests, etc.

Temporary Storage

This step should be as short as possible and better completely eliminated. Vegetables can be stored in:

- Simple stores, without artificial cooling;
- In refrigerated stores; or, in some cases,
- In silos (potatoes, etc.).

Simple stores should be covered, fairly cool, dry and well ventilated but without forced air circulation which can induce significant losses in weight through intensive water evaporation; air relative humidity should be at about 70-80per cent. Refrigerated storage is always preferable and in all cases a processing centre needs a cold room for this purpose, adapted in volume I capacity to the types and quantities of vegetables (and fruits) that are further processed.

Washing

Washing is used not only to remove field soil and surface micro-organisms but also to remove fungicides, insecticides and other pesticides, since there are laws specifying maximum levels of these materials that may be retained on the vegetable; and in most cases the allowable residual level is virtually zero. Washing water contains detergents or other sanitisers that can essentially completely remove these residues.

The washing equipment, like all equipment subsequently used, will depend upon the size, shape and fragility of the particular kind of vegetable:

- Flotation cleaner for peas and other small vegetables;
- Rotary washer in which vegetables are tumbled while they are sprayed with jets of water; this type of washer should not be used to clean fragile vegetables;

Sorting

This step covers two separate operations:

- Removal of non-standard vegetables (and fruit) and possible foreign bodies remaining after washing;
- Quality grading based on variety, dimensional, organoleptical and maturity stage criterion.

Skin Removal/Peeling

Some vegetables require skin removal. This can be done in various ways:

- Mechanical

  This type of operation is performed with various types of equipment which depend upon the result expected and the characteristics of the fruit and vegetables, for example:

  - A machine with abrasion device (potatoes, root vegetables);
  - Equipment with knives (apples, pears, potatoes, etc.);
  - Equipment with rotating sieve drums (root vegetables). Sometimes this operation is simultaneous with washing (potatoes, carrots, etc.) or preceded by blanching (carrots).
- Chemical

  Skins can be softened from the underlying tissues by submerging vegetables in hot alkali solution. Lye may be used at a concentration of about 0.5-3per cent, at about 93° C (2000 F) for a short time period (0.5-3 min). The vegetables with loosened skins are then conveyed under high velocity jets of water which wash away the skins and residual lye. In order to avoid enzymatic browning, this chemical peeling is followed by a short boiling in water or an immersion in diluted citric acid solutions. It is more difficult to peel potatoes with this method because it is necessary to dissolve the cutin and this requires more concentrated lye solutions, up to 10per cent.

- Thermal

Wet Heat (Steam)

Other vegetables with thick skins such as beets, potatoes, carrots and sweet potatoes may be peeled with steam under pressure (about 10 at) as they pass through cylindrical vessels.

This softens the skin and the underlying tissue. When the pressure is suddenly released, steam under the skin expands and causes the skin to puff and crack. The skins are then washed away with jets of water at high pressure (up to 12 at).

Dry Heat (Flame)

Other vegetables such as onions and peppers are best skinned by exposing them to direct flame (about 1 min at 1000° C) or to hot gases in rotary tube flame peelers. Here too, heat causes steam to develop under skins and puff them so that they can be washed away with water.

Table. Losses at Vegetable Peeling, in per cent.

	Peeling methods		
Vegetables	Manual	Mechanical	Chemical
Potatoes	15-19	18-28	-
Carrots	13-15	16-18	8-10
Beets	1416	13-15	9-10

Manual peeling only use when the other methods are impossible or sometimes as a completion of the other three ways. Average losses at this step are given in Table.

Size Reduction

This step is applied according to specific vegetable and processing technology requirements.

Blanching

The special heat treatment to inactivate enzymes is known as blanching. Blanching is not indiscriminate heating. Too little is ineffective, and too much damages the vegetables by excessive cooking, especially where the fresh character of the vegetable is subsequently to be preserved by processing.

This heat treatment is applied according to and depends upon the specificity of vegetables, the objectives that are followed and the subsequent processing/preservation methods.

Two of the more heat resistant enzymes important in vegetables are catalase and peroxidase. If these are destroyed then the other significant enzymes in vegetables also will have been inactivated. The heat treatment to destroy catalase and peroxidase in different vegetables are known, and sensitive chemical tests have been developed to detect the amounts of these enzymes that might survive a blanching treatment.

Because various types of vegetables differ in size, shape, heat conductivity, and the natural levels of their enzymes, blanching treatments have to be established on an experimental basis. As with sterilisation of foods in cans, the larger the food item the longer it takes for heat to reach the centre. Small vegetables may be adequately blanched in boiling water in a minute or two, large vegetables may require several minutes.

Blanching as a unit operation is a short time heating in water at temperatures of 100° C or below. Water blanching may be performed in double bottom kettles, in special baths with conveyor belts or in modern continuous blanching equipment.

In order to reduce losses of hydrosoluble substances (mineral salts, vitamins, sugars, etc.) occurring during water blanching, several methods have been developed:

- Temperature setting at 85-95° C instead of 100° C;
- Blanching time has to be just sufficient to inactivate enzymes catalase and peroxidase;
- Assure elimination of air from tissues.

An illustration of blanching parametres is seen in Table given below.

Table. Blanching Parametres for Some Vegetables.

Vegetables	Temperature, °C	Time, min.
Peas	85-90	2-7
Green beans	90-95	2-5
Cauliflower	Boiling	2
Carrots	90	3-5
Peppers	90	3

Steam heat treatment can also be applied instead of water blanching as a preliminary step before freezing or drying, as long as the preservation method is only used for enzyme inactivation and not to modify consistency. For drying, the vegetables are conveyed directly from steaming equipment to drying installations without cooling. Vegetable steaming is carried out in continuous installations with conveyer belts made from metallic sieves.

Cooling of vegetables after water blanching or steaming is performed in order to avoid excessive softening of the tissues and has to follow immediately after these operations; one exception is the case of vegetables for drying which can be transferred directly to drying equipment without cooling.

Natural cooling is not recommended because is too long and generates significant losses in vitamin C content. Cooling in pre-cooled air (from special installations) is sometimes used for vegetables that will be frozen

Cooling in water can be achieved by sprays or by immersion; in any case the vegetables have to reach a temperature value under 37° C as soon as possible. Too long a cooling time generates supplementary losses in valuable hydrosoluble substances; in order to avoid this, the temperature of the cooling water has to be as low as possible.

Canning

Large quantities of vegetable products are canned. A typical flow sheet for a vegetable canning operation (which also applies to fruit for the most part) covers some food process unit operations performed in sequence: harvesting; receiving; washing; grading; heat blanching; peeling and coring; can filling; removal of air under vacuum; sealing/closing, retorting/ heat treatment; cooling; labelling and packing. The vegetable may be canned whole, diced, purḗed, as juice and so on.

On-line Simplified Methods for Enzyme Activity Check

Peroxidase test:

- Solutions. In order to check the peroxidase activity two solutions have to be prepared:
  - 1per cent guaiacol in alcohol solution (1 g guaiacol is dissolved in about 50 cm³ of 96per cent ethylic alcohol and then this preparation is brought to 100 cm° with the same solvent);
  - Peroxide solution 0.3per cent (1 cm³ perhydrol is brought to 100 cm³ with distilled water.
- Sampling. From various parts of the material samples are taken (about 20-30 pieces, etc.); the material is then crushed in a labouratory bowl in order to obtain an average sample.
- Check. Prom the average sample, 10-20 g of material is taken in a medium capacity test tube; on this sample are poured: 20 cm³ distilled water; 1 cm³ of 1per cent guaiacol solution; 1.6 cm³ of peroxide solution.

The contents of the test tube is shaken well. The gradual appearance of a weak pink colour indicates an incomplete peroxidase inactivation - reaction slightly positive. If there are no tissue colour modifications after 5 minutes, the reaction is negative and the enzymes have been inactivated.

As an orientative check it is also possible to simply pour a few drops of 1per cent guaiacol solution and 0.3per cent peroxide solution directly on blanched and crushed vegetables. A rapid and intensive brown-reddish tissue colouring indicates a high peroxidase activity (positive reaction).

Catalase Test

In order to identify the catalase enzyme activity, 2 g of dehydrated vegetables are well crushed and mixed with about 20 cm^3 of distilled water. After 15 min softening, 0.5 cm^3 of a 0.5per cent or 1per cent peroxide solution is poured on prepared vegetables. In the presence of catalase, a strong oxygen generation is observed for about 2-3 minutes.

These tests are of a paramount importance in order to determine the vegetable blanching treatments (temperature and time); incomplete enzyme inactivation has a negative effect on finished product quality.

For cabbage catalase inactivation by blanching is sufficient; blanching further to peroxidase inactivation would have negative effects on product quality and even complete browning. For all other vegetables and for potatoes, both tests MUST be negative, for catalase and for peroxidase.

Fresh Vegetable Storage

The vegetables can be stored, in some specific natural conditions, in fresh state, that is without significant modifications of their initial organoleptic properties. Fresh vegetable storage can be short term; this was briefly covered under temporary storage before processing. Also fresh vegetable storage can be long term during the cold season in some countries and in this case it is an important method for vegetable preservation in the natural state. In order to assure preservation in long term storage, it is necessary to reduce respiration and transpiration intensity to a minimum possible; this can be achieved by:

- Maintenance of as low a temperature as possible (down to 0° C),
- Air relative humidity increased up to 85-95 per cent and
- CO_2 percentage in air related to the vegetable species.

Vegetables for storage must conform to following conditions: they must be of one of the autumn or winter type variety; be at edible maturity without going past this stage; be harvested during dry days; be protected from rain, sun heat or wind; be in a sound state and clean from soil; be undamaged.

From the time of harvest and during all the period of their storage vegetables are subject to respiration and transpiration and this is on account of their reserve substances and water content. The more the intensity of these two natural processes are reduced, the longer sound storage time will be and the more losses will be reduced. For this reason, vegetables have to be handled and transported as soon as possible in the storage conditions (optimal temperature and air relative humidity for the given species). Even in these optimal conditions storage will generate losses in weight which are variable and depend upon the species.

Vegetable Natural Acidification Technology

Raw materials must follow strict specifications for a high quality finished product; the following parametres must be considered as critical:

- Adapt a uniform size according to the finished product requirements; for example, gherkins will need to have a maximum length of 9 cm for raw vegetables. Generally 15 cm size/length will be a maximum for high quality cucumber products in many countries. However, according to local preferences, bigger cucumbers could be also in demand.
- Cylindrical or ovoidal shape;
- Dark green colour;
- Absence of surface defects due to cryptogamic diseases.

Cucumbers have to be picked at their ripeness for eating, when the sugar content is at about 1.5-2.2per cent, needed for lactic fermentation. Unripe cucumber does not have enough sugar.

- Small holes are made in large size cucumbers skin;
- *Rreceptacle filling:* raw material is simply put in the receptacles in bulk, with care to arrange them in such a way that a maximum of pieces could be introduced;
- *Salt solution preparation:* 6per cent salt solution (NaCl);
- *Salt solution addition*: the salt solution is poured into the receptacle;
- Fermentation is carried out at 20-30° C, anaerobically. This step takes generally 4-8 weeks. Acidity reaches a value up to 1.5per cent lactic acid (and in some exceptional cases up to 2per cent lactic acid) which corresponds to a maximum pH value of 4.1.
- Storage; after the last fermentation stage, drums and other receptacles have to be stored at low temperature; best conditions for 12 months shelf life should be below + 15° C. Storage temperature will determine the shelf life of the products.
- Addition of 1000 ppm potassium sorbate will prevent mould development without having any influence on lactic fermentation.
- Raw material grading by size is a very important technological step. In order to accelerate brine penetration, mainly for medium

to large size cucumbers, the practice of making small holes in the raw material skins is generally recommended.

- A major factor influencing the quality of lactic fermented cucumbers is the water durity; optimal results are obtained at 15-20° durity.
- Cucumber consistency/texture is influenced by the formation of calcium pectate with the pectic substances from raw material tissues. In some countries, calcium chloride (0.3-0.5 per cent) is added in order to firm up the cucumber consistency. Chlorinated water which still contains active chlorine can inhibit or even stop the lactic fermentation.

Sauerkraut

In some countries cabbages are submitted to lactic fermentation as whole vegetables; however, in many countries the cabbage is shredded before fermentation.

As shredded cabbage and its technology is at the basis of an important industry, giving good quality products, with a uniform fermented product and with good keeping quality and ease of distribution, this will be described first.

Cabbage as raw material for sauerkraut must be sound, ripe for eating, well-leafed and from suitable varieties.

Optimum total sugar level needed for the lactic fermentation is 24per cent; generally good quality raw material contains up to 30-60 mg/100 g of vitamin C.

Removal of External Leaves

Coring is done with a specially adapted mechanical screw; this operation generates small particles of finely divided cabbage which will be mixed with the main part of vegetable during shredding/chopping.

The core represents about 10per cent from the whole cabbage, is rich in sugar and vitamin C, but being too high in fibre content needs to be chopped separately as described.

Shredding/Cutting of cabbage is carried out with complex specific equipment which is generally installed directly on the "top" of fermentation silos and is mobile, installed on rails and moves all along the silos. The dimension of resulting shredded cabbage is about 2-3 mm thick.

The same complex equipment is designed to grind the added salt to fine particles and to distribute shredded cabbage and ground salt in an uniform manner to the fermentation silos.

The usual capacity of fermentation silos is up to 30 tons, with separate compartments of 45 tons each.

Salt addition is carried out by the equipment described above; the proportion of salt is 2-2.5per cent with respect to cabbage.

This proportion must not be changed because the salt in this technology does not have a preservative role but only that to extract from cabbage the juice needed for fermentation. It is be preferable to obtain a fairly light pressure on cabbage just after salt addition with some simple mechanical means.

This is important in order to:

- Create an anaerobic medium for fermentation;
- Facilitate external diffusion of cellular juice;
- Assure a rational use of the fermentation space.

Fermentation: The maximum acidity level obtained is generally of about 1.5per cent lactic acid (and very rarely 2.5per cent); this is obtained in 4-6 weeks. Optimal acidity is 1.0-1.8per cent and pH value 4.1 or lower.

Fermentation temperature is at 20-25° C in the first phase and needs to be lowered then to 14-18° C.

During fermentation, the brine from each storage/fermentation silo cell is periodically circulated with a pump in order to uniformise the fermentation process.

Storage is performed in same silos used for fermentation, or the finished products is removed from silos and packed in drums and other receptacles according to distribution schedule.

These silos are usually made of reinforced concrete and coated with gritstone plates or with an acid-resisting material layer. At small scale and in traditional processing, shredded sauerkraut can be obtained by using simple available glass or rigid plastic receptacles.

At home, this process can use glass jars and/or local/traditional pottery receptacles from a minimum size of 2-3 kg up to the available/practical sizes (better limited to 10-15 kg).

In some countries shredded sauerkraut is preserved in receptacles by pasteurization, once the fermentation process has been completed.

Whole Sauerkraut

According to the consumer preference in different countries and to the specific situations it is also usual to preserve whole cabbages by lactic fermentation. At small or medium scale operations, whole cabbage could be processed/fermented in cylindric receptacles like 30 to 200 litre rigid plastic drums, or rectangular receptacles made from food grade rigid plastic.

It is possible to find this type of drum in a significant number of developing countries. These two types of rigid plastic receptacles could also be used for shredded sauerkraut production. Prepared whole cabbages are put into fermentation receptacles and a 5-6 per cent salt concentration brine is poured on top.

The fermentation conditions are the same as for shredded sauerkraut. In order to assure a uniform fermentation and to avoid a strict anaerobic (butyric) fermentation it is necessary to apply a periodic juice "aeration" (each 2-3 days at the beginning of the fermentation, and then each 5-7 days).

Other Acidified Vegetables

In principle all vegetables with a sugar content of at least 2 per cent could be preserved by lactic fermentation. From a practical point of view it is mainly the following vegetables which are preserved by this technology: unripe tomatoes (green tomatoes), peppers, eggplant, carrots and cauliflower, alone or usually in a mix with cucumber as mixed pickles.

Fermentation of individual vegetables is carried out according to a flow-sheet as described for whole sauerkraut. The type of cut, brine concentration and frequency of operating steps have to be adapted to each case; green tomatoes are fermented as whole vegetable.

5

Preservation of Fruits and Vegetables

Fruit preserves

Fruit preserves are fruits, or vegetables, that have been prepared and canned or sealed air-tight for long term storage. The preparation of fruit preserves traditionally involves the use of pectin as a gelling agent, although sugar or honey may be used as well. The ingredients used and how they are prepared will determine the type of preserves; jams, jellies and marmalades are all examples of different styles of fruit preserves that vary based upon the ingredients used.

There are various varieties of fruit preserves made globally, and they can be made from sweet or savory ingredients. In North America, the plural form preserves is used, while the singular preserve is used in British and Commonwealth English. Additionally, the name of the type of fruit preserves will also vary depending on the regional variant of English being used.

Variations

Confit

Confit, which is the past participle form of the French verb "confire" or "to preserve", is most often applied to preservation of meats, especially poultry and pork, by cooking them in their own fat or oils and allowing the fats to set. However, the term can also refer to fruit or vegetables which have been seasoned and cooked with honey or sugar until it has reached a jam-like consistency. Savory confits, such as ones made with garlic or tomatoes, may call for a savory oil such as virgin olive oil as the preserving agent.

Conserves

A conserve is a jam made of fruit stewed in sugar. Often the making of conserves can be trickier than making a standard jam, because the balance between cooking, or sometimes steeping in the hot sugar mixture for just enough time to allow the flavour to be extracted from the fruit, and sugar to penetrate the fruit, and cooking too long that fruit will break down and liquefy.

This process can also be achieved by spreading the dry sugar over raw fruit in layers, and leaving for several hours to steep into the fruit then just heating the resulting mixture only to bring to the setting point. As a result of this minimal cooking, some fruits are not particularly suitable for making into conserves, because they require cooking for longer periods to avoid issues such as tough skins. Currants and gooseberries, and a number of plums are among these fruits. Due to this shorter cooking period, not as much pectin will be released from the fruit, and as such, conserves will sometimes be slightly softer set than some jams. An alternate definition holds that conserves are preserves made from a mixture of fruits and/or vegetables. Conserves may also include dried fruit or nuts.

Fruit Butter

Fruit butter, in this context, refers to a process where the whole fruit is forced through a sieve or blended after the heating process:

- "Fruit butters are generally made from larger fruits, such as apples, plums, peaches or grapes. Cook until softened and run through a sieve to give a smooth consistency. After sieving, cook the pulp... add sugar and cook as rapidly as possible with constant stirring.... The finished product should mound up when dropped from a spoon, but should not cut like jelly. Neither should there be any free liquid."

Fruit Curd

Fruit curd is a dessert topping and spread usually made with lemon, lime, orange, or raspberry. The basic ingredients are beaten egg yolks, sugar, fruit juice and zest which are gently cooked together until thick and then allowed to cool, forming a soft, smooth, intensely flavoured spread. Some recipes also include egg whites and/or butter.

Fruit Spread

Fruit spread refers to a jam or preserve with no added sugar.

Jam

Jam contains both fruit juice and pieces of the fruit's flesh, however, some cookbooks define jam as cooked and gelled fruit purees. Properly, the term jam refers to a product made with whole fruit, cut into pieces or crushed. The fruit is heated with water and sugar to activate the pectin in the fruit. The mixture is then put into containers. The following extract from a US cookbook describes the process:

- "Jams are usually made from pulp and juice of one fruit, rather than a combination of several fruits. Berries and other small fruits are most frequently used, though larger fruits such as apricots, peaches, or plums cut into small pieces or crushed are also used for jams. Good jam has a soft even consistency without distinct pieces of fruit, a bright colour, a good fruit flavour and a semi-jellied texture that is easy to spread but has no free liquid."

Jelly

Jelly is a clear or translucent fruit spread made from sweetened fruit juice and set using naturally occurring pectin. Additional pectin may be added where the original fruit does not supply enough, for example with grapes. Jelly can be made from sweet, savory or hot ingredients. It is made by a process similar to that used for making jam, with the additional step of filtering out the fruit pulp after the initial heating. A muslin or stockinette "jelly bag" is traditionally used as a filter, suspended by string over a bowl to allow the straining to occur gently under gravity. It is important not to attempt to force the straining process, for example by squeezing the mass of fruit in the muslin, or the clarity of the resulting jelly will be compromised.

- "Good jelly is clear and sparkling and has a fresh flavour of the fruit from which it is made. It is tender enough to quiver when moved, but holds angles when cut.
- *Extracting Juice*: Pectin is best extracted from the fruit by heat, therefore cook the fruit until soft before straining to obtain the juice.... Pour cooked fruit into a jelly bag which has been wrung out of cold water. Hang up and let drain. When dripping has ceased the bag may be squeezed to remove remaining juice, but this may cause cloudy jelly."

Marmalade

British-style marmalade is a sweet preserve with a bitter tang made from fruit, sugar, water, and a gelling agent. American-style marmalade is sweet, not bitter. In English-speaking usage, "marmalade" almost always refers to a preserve derived from a citrus fruit, most commonly oranges, although onion marmalade is also used as an accompaniment to savoury dishes. The recipe includes sliced or chopped fruit peel, which is simmered in fruit juice and water until soft; indeed marmalade is sometimes described as jam with fruit peel. Such marmalade is most often consumed on toasted bread for breakfast.

The favoured citrus fruit for marmalade production in the UK is the "Seville orange", *Citrus aurantium* var. aurantium, thus called because it was originally imported from Seville in Spain; it is higher in pectin than sweet oranges, and therefore gives a good set. Marmalade can also be made from lemons, limes, grapefruit, strawberries or a combination.

Regional terminology

The term preserves is usually interchangeable with jam. Some cookbooks define preserves as cooked and gelled whole fruit which includes a significant portion of the fruit. The terms jam and jelly are used in different parts of the English-speaking world in different ways. In the United States, both jam and jelly are sometimes popularly referred to as "jelly", whereas in the United Kingdom, Canada, India and Australia, the two terms are more strictly differentiated. In Australia and South Africa, the term jam is more popularly used as a generic term for both jam and jelly. To further confuse the issue, the term jelly is also used in the UK, South Africa, Australia, India and New Zealand to refer to a gelatin dessert, known in North America as jello, derived from the brand name Jell-O.

Production

In general, jam is produced by taking mashed or chopped fruit or vegetable pulp and boiling it with sugar and water. The proportion of sugar and fruit varies just as to the type of fruit and its ripeness, but a rough starting point is equal weights of each. When the mixture reaches a temperature of 104°C, the acid and the pectin in the fruit react with the sugar, and the jam will set on cooling. However, most cooks work by trial and error, bringing the mixture to a "fast rolling boil", watching to see if the seething mass changes texture, and dropping small samples on a plate to see if they run or set. Commercially produced jams are usually produced using one of two methods. The first is the open pan method, which is essentially a larger scale version of the method a home jam maker would use. This gives a traditional flavour, with some caramelization of the sugars.

The second commercial process involves the use of a vacuum vessel, where the jam is placed under a vacuum, which has the effect of reducing its boiling temperature to anywhere between 65-80°C depending on the recipe and the end result desired.

The lower boiling temperature enables the water to be driven off as it would be when using the traditional open pan method, but with the added benefit of retaining more of the volatile flavour compounds from the fruit, preventing caramelization of the sugars, and of course reducing the overall energy required to make the product. However, once the desired amount of water has been driven off, the jam still needs to be heated briefly to 95-100 °C to kill off any micro-organisms that may be present; the vacuum pan method does not kill them all.

During the commercial filling of the jam into jars, it is common to use a flame to sterilize the rim of the jar and the lid to destroy any yeasts and molds which may cause spoilage during storage. It is also common practice to inject steam into the head space at the top of the jar immediately prior to the fitting of the lid, in order to create a vacuum. Not only does this vacuum help prevent the growth of spoilage organisms, it also pulls down the tamper evident safety button when lids of this type are employed. How easily a jam sets depends on the pectin content of the fruit.

Some fruits, such as gooseberries, redcurrants, blackcurrants, most citrus fruits, apples, and raspberries, set very well; others, such as strawberries and ripe blackberries, often need to have pectin added. There are commercial pectin products on the market, and most industrially-produced jams use them. Home jam-makers sometimes rely on adding a pectin-rich fruit to a poor setter, for example apple to blackberries. Other tricks include extracting juice from lemons, redcurrants or gooseberries, or making a pectin stock with whole apples or just the cores and skins; once cooled, this "stock" can then be frozen for later use. Making jam at home is a popular handicraft activity, and many take part in this. Homemade jam may be made for personal consumption, or as part of a cottage industry.

Aspic

Aspic is a dish in which ingredients are set into a gelatin made from a meat stock or consommé. Similar dishes, made with commercial gelatin mixes instead of stock or consommé, are usually called gelatin salads. When cooled, stock that is made from meat congeals because of the natural gelatin found in the meat. The stock can be clarified with egg whites, and then filled and flavoured just before the aspic sets. Almost any type of food can be set into aspics.

Most common are meat pieces, fruits, or vegetables. Aspics are usually served on cold plates so that the gel will not melt before being eaten. A meat jelly that includes cream is called a chaud-froid. Nearly any type of meat can be used to make the gelatin: pork, beef, veal, chicken, turkey, or fish.

Gelatin is also found in cartilage. The aspic may need additional gelatin in order to set properly. Veal stock provides a great deal of gelatin; in making stock, veal is often included with other meat for that reason. Fish consommés usually have too little natural gelatin, so the fish stock may be double-cooked or supplemented. Since fish gelatin melts at a lower temperature than gelatins of other meats, fish aspic is more delicate and melts more readily in the mouth. Vegetables and fish stocks need gelatin to create a mould.

Historically, meat jellies were made before fruit and vegetable jellies. By the Middle Ages at the latest, cooks had discovered that a thickened meat broth could be made into a jelly. A detailed recipe for aspic is found in Le Viandier, written in around 1375. In the 18th century, Marie-Antoine Carême created chaud froid in France. Chaud froid means "hot cold" in French, referring to foods that were prepared hot and served cold. Aspic was used an chaud froid sauce in many cold fish and poultry meals. The sauce added moisture and flavour to the food. Marie-Antoine Carême also invented various types of aspic and ways of preparing aspic.

Aspic didn't come into prominence until the early 20th century. Aspic was used to hold meats and prevented them from becoming spoiled. The gelatin prevented air and bacteria from ruining cooked meat. It wasn't eaten until the 1950s, when it became popular dish. In the 1950s, aspic was a popular dinner staple throughout the United States. Jello was popular in the 1950s and the demands for meat were abundant. One of the most popular dishes were the tomato aspic. Cooks used to show off aesthetic skills by creating inventive aspics.

Uses

Aspic can also be referred as a aspic gelèe or aspic jelly. Aspic jelly may be colourless or contain various shades of amber. Aspic can be used to protect food from the air, give food more flavour, or as a decoration. There are three types of aspic textures: delicate, sliceable, and inedible.

The delicate aspic is soft. The sliceable aspic must be made in a terrine or in a aspic mould. It is more firmer than the delicate aspic. The inedible aspic is never for consumption. It is usually for decoration. Aspic is often used to glaze food pieces in food competitions to make the food glisten and making it more appealing to the eye. Foods dipped in aspic can have a lacquered finish for a fancy presentation. Aspic can be cut into various shapes and be used as a garnish for deli meats or pâtés.

Jugging

In cooking, jugging is the process of stewing meat for a long time in a tightly covered container, such as a casserole or an earthenware jug. Sometimes the cooking liquid includes some of the animal's blood. In French, such a stew of a game animal, thickened with the animal's blood, is known as a civet. One common traditional dish that involves jugging is Jugged Hare which is a whole hare, cut into pieces, marinated and cooked with red wine and juniper berries in a tall jug that stands in a pan of water. It is traditionally served with the hare's blood and Port wine.

The recipe goes on to describe cooking the pieces of hare in water in a jug that is set within a bath of boiling water to cook for three hours.

Beginning in the nineteenth century, Glasse has been widely credited with having started the recipe with the words "First, catch your hare". This attribution is apocryphal. However, having a freshly caught, or shot, hare enables one to obtain its blood.

A freshly killed hare is prepared for jugging by removing its entrails and then hanging it in a larder by its hind legs, which causes the blood to accumulate in the chest cavity. One method of preserving the blood after draining it from the hare is to mix it with red wine vinegar in order to prevent it coagulating, and then to store it in a freezer.

Many other British cookbooks from before the middle of the 20th century have recipes for Jugged Hare. Merle and Reitch have this to say about Jugged Hare, for example:

- The best part of the hare, when roasted, is the loin and the thick part of the hind leg; the other parts are only fit for stewing, hashing, or jugging. It is usual to roast a hare first, and to stew or jug the portion which is not eaten the first day. To Jug a Hare: This mode of cooking a hare is very desirable when there is any doubt as to its age, as an old hare, which would be otherwise uneatable, may be made into an agreeable dish. In 2006, a survey of 2021 people for the television channel UKTV Food found that only 1.6 per cent of the people aged under 25 recognized Jugged Hare by name. 7 out of 10 of those people stated that they would refuse to eat Jugged Hare if it were served at the house of a friend or a relative. Jugged Rabbit is an alternative to Jugged Hare. It is considered a speciality of the cuisine of Martinique. Another jugged dish, also traditional in the United Kingdom, is Jugged Kippers, which is kippers in a covered jug, cooked in boiling water.

Modified atmosphere

Modified atmosphere is an uncommon technical definition that describes the practice of modifying the composition of the internal atmosphere of a package in order to improve the shelf life.

The modification process often tries to lower the amount of oxygen, moving it from 20 per cent to 0 per cent, In order to slow down the growth of aerobic organisms and the speed of oxidation reactions.

The removed oxygen can be replaced with nitrogen, commonly acknowledged as an inert gas, or carbon dioxide, which can lower the pH or inhibit the growth of bacteria. Carbon monoxide can be used for keeping the red colour of meat. Re-balancing of gases inside the packaging can be achieved using active techniques such as gas flushing and compensated vacuum or passively by designing "breathable" films known as equilibrium modified atmosphere packaging.

Scientific terms

- MAP = Modified atmosphere packaging
- EMAP = Equilibrium modified atmosphere packaging
- MA/MH = Modified atmosphere/modified humidity packaging.

Products

Under MA products like red meat, seafood, minimally processed fruits and vegetables, pasta, cheese, bakery goods, poultry, cooked and cured meats, ready meals and dried foods are packaged. The three major commodity types are fruits and vegetables, meat and meat products and seafood. It has been estimated that 25-40 per cent of all fresh produce harvested will not reach the consumers table, due to spoilage and mishandling that occurs during distribution.

Modified Atmosphere Packaging

Modified Atmosphere Packaging is a technique used for prolonging the shelf-life period of fresh or minimally processed foods. In this preservation technique the air surrounding the food in the package is changed to another composition.

This way the initial fresh state of the product may be prolonged. It is the shelf-life of perishable products like meat, fish, fruits and vegetables that will be prolonged with MAP since it slows the natural deterioration of the product. MAP is used with various types of products, where the mixture of gases In the package depends on the type of product, packaging materials and storage temperature.

Meat and fish need very low gas permeability films so for non-respiring products high barrier films are used. The initial flushed gas-mixture will be maintained inside the MA package. But fruits and vegetables are respiring products where the interaction of the packaging material with the product is important. If the permeability of the packaging film is adapted to the products respiration, an equilibrium modified atmosphere will establish in the package and the shelf-life of the product will increase.

Equilibrium modified atmosphere packaging

Among fresh-cut produce Equilibrium Modified Atmosphere Packaging is the most commonly used packaging technology. When packaging vegetables and fruits the gas atmosphere of package is not air but consists usually of a lowered level of O_2 and a heightened level of CO_2.

This kind of package slows down the normal respiration of the product and so to prolong the shelf-life of the product. Of course there are other factors like the size of the product, severity of preparation, maturity of the product and type of tissue that have an effect to the shelf-life of an EMA packaged produce.

Technology

There are two techniques used in the industry to pack vegetables namely gas-flushing and compensated vacuum. For its cheapness the gas-flushing is more widely used. In gas-flushing the package is flushed with a desired gas mixture, as in compensated vacuum the air is removed totally and the desired gas mixture then inserted.

The label "packaged in a protective atmosphere" can refer to either of these; an example of a gas mixture used for non-vegetable packaged food is 99.9 per cent nitrogen gas, which is inert at the temperatures and pressures the packaging is subjected to.

Gases

The atmosphere in an MA package consists of N_2, O_2, CO_2. It is the altered ratio of these gases that makes a difference in the prolongation of shelf life. By reducing the O_2-level and increasing the CO_2-level, ripening of fruits and vegetables can be delayed, respiration and ethylene production rates can be reduced, softening can be retarded and various compositional changes associated with ripening can be slowed down. Oxygen is essential when packaging fresh fruits and vegetables as they continue to respire after harvesting.

The absence of O_2 can lead to anaerobic respiration in the package which accelerates senescence and spoilage. Too high levels of O_2 do not retard respiration significantly and it is around 12 per cent of O_2 where the respiration rate starts to decrease. So oxygen is used in low levels for positive effect. When packaging meat and fish, the high CO_2-levels are effective bacterial and fungal growth inhibitors.

In the case of vegetables and fruits, CO_2 is not a major factor since CO_2-levels above 10 per cent are needed to suppress fungal growth significantly. Unfortunately higher levels than 10 per cent of CO_2 are working phytotoxic for fresh produce. Nitrogen is used as a filler gas since it neither encourages or discourages bacterial growth.

Packaging Films

When selecting packaging films for EMAP of fruits and vegetables the main characteristics to consider are gas permeability, water vapour transmission rate, mechanical properties, transparency, type of package and sealing reliability. Traditionally used packaging films like LDPE, PVC, EVA and OPP are not permeable enough for highly respiring products like fresh-cut produces, mushrooms and broccoli. As fruits and vegetables are respiring products, there is a need to transmit gases from and to the package. Films designed with these properties are called permeable films. Other films, called barrier films, are designed to prevent the exchange of gases and are mainly used with non-respiring products like meat and fish. EMAP films developed to control the humidity level as well as the gas composition in the sealed package are beneficial for the prolonged storage of fresh fruits, vegetables and herbs that are sensitive to moisture.

Preservation of vegetables by acidification

Food acidification is a means of preventing their deterioration in so far as a non-favourable medium for micro-organisms development is created. This acidification can be obtained by two ways: natural acidification and artificial acidification.

Natural Acidification

This is achieved by a predominant lactic fermentation which assures the preservation based on acidoceno-anabiosys principle; preservation by lactic fermentation is called also biochemical preservation. Throughout recorded history food has been preserved by fermentation. In spite of the introduction of modern preservation methods, lactic acid fermented vegetables still enjoy a great popularity, mainly because of their nutritional and gastronomic qualities.

The various preservation methods discussed thus far, based on the application of heat, removal of water, cold and other principles, all have the common objective of decreasing the number of living organisms in foods or at least holding them in check against further multiplication.

Fermentation processes for preservation purposes, in contrast, encourage the multiplication of micro-organisms and their metabolic activities in foods. But the organisms that are encouraged are from a select group and their metabolic activities and end products are highly desirable. The extent of this desirability is emphasised by a partial list of fermented fruits and vegetable products from various parts of the world.

There are some characteristic features in the production of fermented vegetables which will be pointed out below using cucumbers as an example. In the production of lactic acid fermented cucumbers, the raw material is put into a brine without previous heating. Through the effect of salt and oxygen deficiency the cucumber tissues gradually die. At the same time, the semi-permeability of the cell membranes is lost, whereby soluble cell components diffuse into the brine and serve as food substrate for the micro-organisms.

Under such specific conditions of the brine the lactic acid bacteria succeed in overcoming the accompanying micro-organisms and lactic acid as the main metabolic products is formed. Under favourable conditions (for example moderate salt in the brine, use of starter cultures) it takes at least 3 days until the critical pH value of 4.1 or less - desired for microbiological reasons - is reached.

Beside the typical taste, for the consumer a crisp texture is the most important quality criterion for fermented vegetables. Because there is no heating step before the fermentation, the indigenous plant enzymes in the fermenting materials are still present during the very first phase. After the destruction of the cell membranes they easily get to their active sites and under favourable conditions they can easily cause softening.

The environmental conditions act in a different manner on single enzymes or enzymes systems: some enzymes are strongly inhibited by salt, others are activated, and in the acid pH-region many enzymes are irreversibly inactivated. Beside indigenous enzymes also enzymes produced by micro-organisms can be responsible for the undesired soft products. In technically advanced societies the major importance of fermented foods has come to be variety they add to the diet. However, in many less developed areas of the world, fermentation and natural drying are the major food preservation methods and as such are vital to survival of a large proportion of the world's current population.

Artificial acidification is carried out by adding acetic acid which is the only organic acid harmless for human health and stable in specific working conditions; in this case biological principles of the preservation are acidoanabiosys and, to a lesser extent, acidoabiosys.

Combined acidification is a preservation technology which involves as a preliminary processing step a weak lactic fermentation followed by acidification (vinegar addition). The two main classes of vegetables preserved by acidification are sauerkraut and pickles; the definitions of these products adapted from US Code of Federal Register are as follows.

Bulk sauerkraut. Bulk or barrelled sauerkraut is the product of characteristic acid flavour, obtained by the full fermentation, chiefly lactic, of properly prepared and shredded cabbage in the presence of 2-3per cent salt. On completion of fermentation, it contains not more than 1.5per cent of acid, expressed as lactic acid.

Canned sauerkraut. Canned (or packaged) sauerkraut, is prepared from clean, sound, well-matured heads of the cabbage plant (Brassica oleracea var. capitata L.) which have been properly trimmed and cut; to which salt is added and which is cured by natural fermentation. The product may or may not be packed with pickled peppers, pimientos, or tomatoes or contain other flavouring ingredients to give the product specific flavour characteristics. The product:

- May be canned by processing sufficiently by heat to assure preservation in hermetically sealed containers; or
- May be packaged in sealed containers and preserved with or without the addition of benzoate of soda or any other ingredient permissible under the provisions of Food and Drug Administration (FDA).

Pickles. "Pickles" means the product prepared entirely or predominantly from cucumbers (Cucumis sativus L.). Clean, sound ingredients are used which may or may not have been previously subjected to fermentation and curing in a salt brine (solution of sodium chloride, NaCl).

The prepared pickles are packed in a vinegar solution to which may be added salt and other vegetables, nutritive sweeteners, seasonings, flavourings, spices, and other ingredients permissible under FDA regulations. The product is packed in suitable containers and heat treated, or otherwise processed to assure preservation. Sauerkraut and pickle products can be preserved under the effect of natural or added acidity, followed by pasteurization when this acidification is not sufficient. Sauerkraut is a very good source of vitamin C; the importance of this product should be emphasised in developing countries as a simple technology which can be applied mainly for consumption of the finished products in remote, isolated areas during the cold season. It is also a excellent technology to be learned to schools which have their own source of cabbage and cucumbers through school agricultural farms.

Sauerkraut and pickles are manufactured on an industrial scale in significant quantities world-wide. However, the basic technology is simple and could be applied at home, farm and community level after some explanation and training. The natural acidification preservation could be considered similar to sun/solar drying in terms of training and development.

The principle of this technology is to add sugar in a quantity that is necessary to augment the osmotic pressure of the product's liquid phase at a level which will prevent microorganism development. From a practical point of view, however, it is usual to partially remove water (by boiling) from the product to be preserved, with the objective of obtaining a higher sugar concentration. In concentrations of 60per cent in the finished products, the sugar generally assures food preservation. It is important to know the ratio between the total sugar quantity in the finished product and the total sugar concentration in the liquid phase because this determines, in practice, the sugar preserving action. The percent composition of a product preserved with sugar, for example marmalade, can be expressed as follows: [I + S + S + N + W] = 100;

I = insoluble substance;

S = sugar from fruits;

S= added sucrose;

N= soluble "non sugar"

W= water.

In this case, total sugar concentration, in the liquid phase, of the finished product is:

$X = 100 (S + s)/100 - (n + i)$ [per cent]

Therefore, in the case of a standard marmalade with 55 per cent sugar added (calculated on the finished product basis), the real concentration in the liquid phase is for example:

$X = 100 (55 + 8)/100 - (5 + 3) = 68.5$per cent

In the food preservation with sugar, the water activity cannot be reduced below 0.845; this value is sufficient for bacteria and neosmophile yeast inhibition but does not prevent mould attack. For this reason, various means are used to avoid mould development:

- Finished product pasteurization (jams, jellies, etc.);
- Use of chemical preservatives in order to obtain the antiseptisation of the product surface.

It is very important from a practical point of view to avoid any product contamination after boiling and to assure an hygienic operation of the whole technological process (this will contribute to the prevention of product moulding or fermentation).

Storage of the finished products in good conditions can only be achieved by ensuring the above level of water activity.

Methods of Reducing Deterioration

A knowledge of deterioration factors and the way they act, including the rates of deterioration to a specific category of food, means that it is possible to list the ways of lowering or stopping the action and obtaining fruit and vegetable preservation.

In order to maintain their nutritional value and organoleptic properties and because of technical-economical considerations, not all the identified means against deterioration actually have practical applications for fruit and vegetable preservation.

Technical methods of reducing food deterioration

This classification of methods of reducing deterioration presents some difficulties because their preservation effects are physical, physico-chemical, chemical and biochemical complex phenomena which rarely act in isolation. Normally they take place together or one after the other.

From the whole list of possible methods of reducing deterioration, over the years, some procedures for fruit and vegetable preservation have found practical application. These technical means can be summarised as follows:

Physical Cooling	Heating Lowering of water content Drying/dehydration.
Concentration Irradiation	Sterilising filtration Other physical means (high pressure, vacuum, inert gases)
Chemical Smoking Artificial acidification	Salting Sugar additio Ethyl alcohol addition Antiseptic substance action
Biochemical	Lactic fermentation (natural acidification) Alcoholic fermentation

Table. Procedures for Fruit and Vegetable Preservation.

Procedures	Practical applications
Fresh storage	Fruits, vegetables
Cold storage	Fruits, vegetables
Freezing	Fruits, vegetables
Drying/dehydration	Fruits, vegetables
Concentration	Fruit and vegetable juices
Chemical preservation	Fruit semi-processed
Preservation with sugar	Fruit products/preserves
Pasteurization	Fruit and vegetable juices
Sterilisation	Fruits, vegetables
Sterilising filtration	Fruit juices
Irradiation	Fruits, vegetables

These preservation procedures have two main characteristics as far as being applied to all food products is concerned:

- Some of them are applied only to one or some categories of foods; others can be used across the board and thus a wider application (cold storage, freezing, drying/dehydration, sterilisation, etc.);

- Some guarantee food preservation on their own while others require combination with other procedures, either as principal or as auxiliary processes in order to assure preservation (for example smoking has to be preceded by salting).

Combined Preservation Procedures

In practice preservation procedures aim at avoiding microbiological and biochemical deterioration which are the principal forms of deterioration. Even with all recent progress achieved in this field, no single one of these technological procedures applied alone can be considered wholly satisfactory from a microbiological, physico-chemical and organoleptic point of view, even if to a great extent the food value is assured.

Thus, heat sterilisation cannot be applied in order to destroy all micro-organisms present in foods without inducing non desirable modifications. Preservation by dehydration/drying assures microbiological stability but has the drawback of undesirable modifications that appear during storage: vitamin losses, oxidation phenomena, etc.

Starting with these considerations, the actual tendency in food preservation is to study the application of combined preservation procedures, aiming at the realisation of maximum efficiency from a microbiological and biological point of view, with reduction to a minimum of organoleptical degradation and decrease in food value.

The principles of combined preservation procedures are:

- Avoid or reduce secondary (undesirable) effects in efficient procedures for microbiological preservation;
- Avoid qualitative degradation appearing during storage of products preserved by efficient procedures from a microbiological point of view;
- Increase microbiological efficiency of preservation procedures by supplementary means;
- Combine preservation procedures in order to obtain maximum efficiency from a microbiological point of view, by specific action on various types of micro-organisms present;
- Establish combined factors that act simultaneously on bacterial cells.

Research and applications in this direction were followed by microbiological and biochemical way, obtaining a serial of combination of preservation procedures with the possibility of application in industrial practice.

Fresh fruit and vegetable storage can be combined with:

- Storage in controlled atmosphere where carbon dioxide and oxygen levels are monitored, increasing concentration of CO2 and lowering that of oxygen according to fruit species. Excellent results were obtained for pomace fruit; in particular the storage period for apples has been extended. Application of this combined procedure requires airtight storage rooms.
- Storage in an environment containing ethylene oxide; this accelerates ripening in some fruit: tomatoes, bananas, mangoes, etc.

Cold storage can be combined with storage in an environment with added of carbon dioxide, sulphur dioxide, etc. according to the nature of product to be preserved.

Preservation by drying/dehydration can be combined with:- freezing: fresh fruit and vegetables are dehydrated up to the point where their weight is reduced by 50per cent and then they are preserved by freezing.

This procedure (freeze-drying) combines the advantages of drying (reduction of volume and weight) with those of freezing (maintaining vitamins and to a large extent organoleptic properties). A significant advantage of this process is the short drying time in so far as it is not necessary to go beyond the inflexion point of the drying curve. The finished products after defreezing and rehydration/reconstitution are of a better quality compared with products obtained by dehydration alone.

- Cold storage of dried/dehydrated vegetables in order to maintain vitamin C; storage temperature can be varied with storage time and can be at -8° C for a storage time of more than one year, with a relative humidity of 70-75 per cent.
- Packaging under vacuum or in inert gases in order to avoid action of atmospheric oxygen;, mainly for products containing beta-carotene.
- Chemical preservation: a process used intensively for prunes and which has commercial applications is to rehydrate the dried product up to 35 per cent using a bath containing hot 2 per cent potassium sorbate solution. Another possible application of this combined procedure is the initial dehydration up to 35per cent moisture followed by immersion in same bath as explained above; this has the advantage of reducing drying time and producing minimum qualitative degradation. Both applications suppress the dehydrated products reconstitution (rehydration) step before consumption.
- Packaging in the presence of desiccants (calcium oxide, anhydrous

calcium chloride, etc.) in order to reduce water vapour content in the package, especially for powdered products.

Preservation by concentration, carried out by evaporation, is combined with cold storage during warm season for tomato paste (when water content cannot be reduced under the limit needed to inhibit moulds and yeasts, *e.g.* a_w = 0.70...0.75). Chemical preservation is combined with:

- Acidification of food medium (lowering pH);
- Using combined chemical preservatives.

Preservation by lactic fermentation (natural acidification) can be combined with cold storage for pickles in order to prolong storage time or shelf-life.

Preservation with sugar is combined with pasteurization for some preserves having a sugar content below 65per cent.

Fresh storage

Fresh Fruit and Vegetable Storage

Once fruit is harvested, any natural resistance to the action of spoiling micro-organisms is lost. Changes in enzymatic systems of the fruit also occur on harvest which may also accelerate the activity of spoilage organisms. Means that are commonly used to prevent spoilage of fruits must include:

- Care to prevent cutting or bruising of the fruit during picking or handling;
- Refrigeration to minimise growth of micro-organisms and reduce enzyme activity;
- Packaging or storage to control respiration rate and ripening;
- Use of preservatives to kill micro-organisms on the fruit.

A principal economic loss occurring during transportation and/or storage of produce such as fresh fruit is the degradation which occurs between the field and the ultimate destination due to the effect of respiration.

Methods to reduce such degradation are as follows:

- Refrigerate the produce to reduce the rate of respiration;
- Vacuum cooling;
- Reduce the oxygen content of the environment in which the produce is kept to a value not above 5per cent of the atmosphere but above the value at which anaerobic respiration would begin. When the oxygen concentration is reduced within 60 minutes the deterioration is in practice negligible.

The following is a summary of some recent developments in post-harvest technology of fresh fruits and vegetables.

Harvest Maturity

This is particularly important with fruit for export. One recent innovation is the measurement of resonant frequency of the fruit which should enable the grading out of over mature and under-mature fruit before they are packed for export.

Harvest Method

Considerable research is continuing on mechanical harvesting of perishable crops with a view to minimising damage. In fruit trees, controlling their height by use of dwarfing rootstocks, pruning and growth regulating chemicals will lead to easier, cheaper more accurate harvesting.

Handling System

Field packing of various vegetables for export has been carried out for many years. In the last decade or so this has been applied, in selected cases, to a few tropical fruit types. Where this system can be practiced it has considerable economic advantages in saving the cost of building, labour and equipment and can result in lower levels of damage into crops.

Pre-cooling

Little innovation has occurred in crop pre-cooling over the last decade. However high velocity, high humidity forced air systems have continued to be developed and refined.

These are suitable for all types of produce and are relatively simple to build and operate and, while not providing the speed of cooling of a vacuum or hydrocooler, have the flexibility to be used with almost all crops.

Chemicals

There is a very strong health lobby whose objective is to reduce the use of chemicals in agriculture and particularly during the post harvest period. Every year sees the prohibition of the use of commonly used post-harvest chemicals. New ways need to be developed to control post-harvest diseases, pest and sprouting.

Coatings

Slowing down the metabolism of fruit and vegetables by coating them with a material which affects their gaseous exchange is being tested and used commercially on a number of products.

Controlled Environment Transport

Recent innovations in this technique have produced great progress as a result of the development and miniaturisation of equipment to measure carbon dioxide and oxygen. Several companies now offer containers where the levels of these two gases can be controlled very precisely.

Water and Water Activity (a_w) in Foods

Micro-organisms in a healthy growing state may contain in excess of 80per cent water. They get this water from the food in which they grow. If the water is removed from the food it also will transfer out of the bacterial cell and multiplication will stop. Partial drying will be less effective than total drying, though for some micro-organisms partial drying may be quite sufficient to arrest bacterial growth and multiplication. Bacteria and yeasts generally require more moisture than moulds, and so moulds often will be found growing on semi-dry foods where bacteria and yeasts find conditions unfavourable; example are moulds growing on partially dried fruits.

Slight differences in relative humidity in the environment in which the food is kept or in the food package can make great differences in the rate of micro-organism multiplication. Since micro-organisms can live in one part of a food that may differ in moisture and other physical and chemical conditions from the food just millimetres away, we must be concerned with conditions in the "microenvironment" of the micro-organisms. Thus it is common to refer to water conditions in terms of specific activity.

The term "water activity" is related to relative humidity. Relative humidity is defined as the ratio of the partial pressure of water vapour in the air to the vapour pressure of pure water at the same temperature. Relative humidity refers to the atmosphere surrounding a material or solution. Water activity or aw is a property of solutions and is the ratio of vapour pressure of the solution compared with the vapour pressure of pure water at the same temperature. Under equilibrium conditions water activity equals:

a_w = RH/100

When we speak of moisture requirements of micro-organisms we really mean water activity in their immediate environment, whether this be in solution, in a particle of food or at a surface in contact with the atmosphere.

At the usual temperatures permitting microbial growth, most bacteria require a water activity in the range of about 0.90 to 1.00.

Some yeasts and moulds grow slowly at a water activity down to as low as about 0.65.

Qualitatively, water activity is a measure of unbound, free water in a system available to support biological and chemical reactions. Water activity, not absolute water content, is what bacteria, enzymes and chemical reactants encounter and is affected by at the micro-environmental level in food materials.

Two foods with the same water content can have very different a_w values depending upon the degree to which water is free or otherwise bound to food constituents. Fig is a representative water absorption isotherm for a given food at a given temperature. It shows the final moisture content the food will have when it reaches moisture equilibrium with atmospheres of different relative humidities.

Thus, this food, at the temperature for which this absorption isotherm was established, will ultimately attain a moisture content of 20per cent at 75per cent RH (relative humidity). If this food was previously dehydrated to below 20per cent moisture and placed in an atmosphere of 75per cent RH, it would absorb moisture until it reached 20per cent. Conversely, if this food was moistened to greater than 20per cent water and then placed at 75per cent RH, it would lose moisture until it reached the equilibrium value of 20per cent.

Under such conditions some foods may reach moisture equilibrium in the very short time of a few hours, others may require days or even weeks. When a food is in moisture equilibrium with its environment, then the a_w of the food will be quantitatively equal to the RH divided by 100.

Qualitatively, water activity is a measure of free or available water, to be distinguished from unavailable or bound water. These states of water also bear a relationship to the characteristic sigmoid shapes of water absorption isotherm curves of various foods.

Thus, according to theory, most of the water corresponding to the portion of the curve below its first inflection point is believed to be tightly bound water, often referred to as an adsorbed mono-molecular layer of water. Moisture corresponding to the region above this point and up to the curve's second inflection point is thought to exist largely as multi-molecular layers of water less tightly held to food constituent surfaces.

Beyond this second inflection point moisture generally is considered to be largely free water condensed in capillaries and interstices within the food. In this latter portion of the sorption isotherm curve small changes in moisture content result in great changes in a food's a_w. In Fig. given below are illustrated the moisture sorption isotherms for various dried fruits at 25°C.

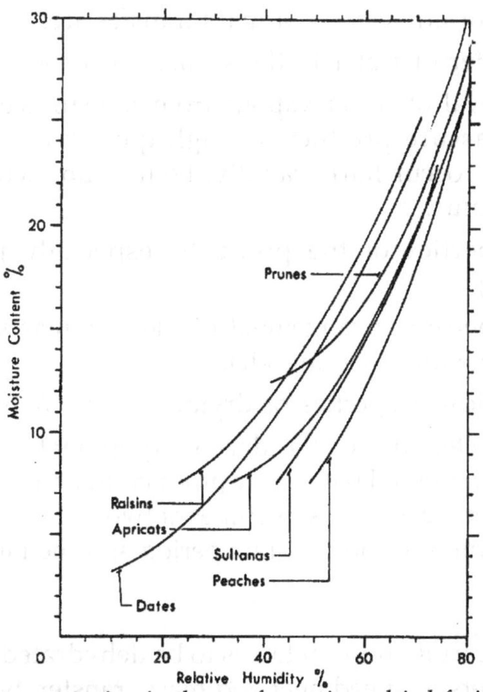

Fig. Moisture sorption isotherms for various dried fruits at 25°C.

Preservation by Drying/Dehydration

The technique of drying is probably the oldest method of food preservation practiced by mankind. The removal of moisture prevents the growth and reproduction of micro-organisms causing decay and minimises many of the moisture mediated deterioration reactions.

It brings about substantial reduction in weight and volume minimising packing, storage and transportation costs and enable storability of the product under ambient temperatures, features especially important for developing countries.

The sharp rise in energy costs has promoted a dramatic upsurge in interest in drying world-wide over the last decade.

Heat and Mass Transfer

Dehydration involves the application of heat to vapourise water and some means of removing water vapour after its separation from the fruit/vegetable tissues.

Hence it is a combined/simultaneous (heat and mass) transfer operation for which energy must be supplied. A current of air is the most common medium for transferring heat to a drying tissue and convection is mainly involved.

The two important aspects of mass transfer are:

- The transfer of water to the surface of material being dried and
- The removal of water vapour from the surface.

In order to assure products of high quality at a reasonable cost, dehydration must occur fairly rapidly. Four main factors affect the rate and total drying time:

- The properties of the products, especially particle size and geometry;
- The geometrical arrangement of the products in relation to heat transfer medium (drying air);
- The physical properties of drying medium/environment;
- The characteristics of the drying equipment.

It is generally observed with many products that the initial rate of drying is constant and then decreases, sometimes at two different rates. The drying curve is divided into the constant rate period and the falling rate period.

Surface Area

Generally the fruit and vegetables to be dehydrated are cut into small pieces or thin layers to speed heat and mass transfer. Subdivision speeds drying for two reasons:

- Large surface areas provide more surface in contact with the heating medium (air) and more surface from which moisture can escape;
- Smaller particles or thinner layers reduce the distance heat must travel to the centre of the food and reduce the distance through which moisture in the centre of the food must travel to reach the surface and escape.

Temperature

The greater the temperature difference between the heating medium and the food the greater will be the rate of heat transfer into the food, which provides the driving force for moisture removal. When the heating medium is air, temperature plays a second important role. As water is driven from the food in the form of water vapour it must be carried away, or else the moisture will create a saturated atmosphere at the food's surface which will slow down the rate of subsequent water removal.

The hotter the air the more moisture it will hold before becoming saturated. Thus, high temperature air in the vicinity of the dehydrating food will take up the moisture being driven from the food to a greater extent than will cooler air. Obviously, a greater volume of air also can take up more moisture than a lesser volume of air.

Air Velocity

Not only will heated air take up more moisture than cool air, but air in motion will be still more effective. Air in motion, that is, high velocity air, in addition to taking up moisture will sweep it away from the drying food's surface, preventing the moisture from creating a saturated atmosphere which would slow down subsequent moisture removal. This is why clothes dry more rapidly on a windy day. Some other phenomena influence the drying process and a few elements are summarised below.

Dryness of Air

When air is the drying medium of food, the drier the air the more rapid is the rate of drying. Dry air is capable of absorbing and holding moisture. Moist air is closer to saturation and so can absorb and hold less additional moisture than if it were dry. But the dryness of the air also determines how low a moisture content the food product can be dried to.

Atmospheric Pressure and Vacuum

If food is placed in a heated vacuum chamber the moisture can be removed from the food at a lower temperature than without a vacuum. Alternatively, for a given temperature, with or without vacuum, the rate of water removal from the food will be greater in the vacuum. Lower drying temperatures and shorter drying times are especially important in the case of heat-sensitive foods.

Evaporation and Temperature

As water evaporates from a surface it cools the surface. The cooling is largely the result of absorption by the water of the latent heat of phase change from liquid to gas. In doing this the heat is taken from the drying air or the heating surface and from the hot food, and so the food piece or droplet is cooled.

Time and Temperature

Since all important methods of food dehydration employ heat, and food constituents are sensitive to heat, compromises must be made between maximum possible drying rate and maintenance of food quality. As is the case in the use of heat for pasteurization and sterilisation, with few exceptions drying processes which employ high temperatures for short times do less damage to food than drying processes employing lower temperatures for longer times.

Thus, vegetable pieces dried in a properly designed oven in four hours would retain greater quality than the same products sun dried over two days. Several drying processes will achieve dehydration in a matter of minutes or even less if the food is sufficiently subdivided.

Drying Techniques

Several types of dryers and drying methods, each better suited for a particular situation, are commercially used to remove moisture from a wide variety of food products including fruit and vegetables.

While sun drying of fruit crops is still practiced for certain fruit such as prunes, figs, apricots, grapes and dates, atmospheric dehydration processes are used for apples, prunes, and several vegetables; continuous processes as tunnel, belt trough, fluidised bed and foam-mat drying are mainly used for vegetables.

Spray drying is suitable for fruit juice concentrates and vacuum dehydration processes are useful for low moisture/high sugar fruits like peaches, pears and apricots. Factors on which the selection of a particular dryer/drying method depends include:

- Form of raw material and its properties;
- Desired physical form and characteristics of dried product;
- Necessary operating conditions;
- Operation costs.

There are three basic types of drying process:

- Sun drying and solar drying;
- Atmospheric drying including batch (kiln, tower and cabinet dryers) and continuous (tunnel, belt, belt-trough, fluidised bed, explosion puff, foam-mat, spray, drum and microwave);
- Sub-atmospheric dehydration (vacuum shelf/belt/drum and freeze dryers).

The scope has been expanded to include use of low temperature, low energy process like osmotic dehydration. As far dryers are concerned, one useful division of dryer types separates them into air convection dryers, drum or roller dryers, and vacuum dryers. Using this breakdown, Table given below indicates the applicability of the more common dryer types to liquid and solid type foods.

Table. Common Dryer Types Used for Liquid and Solid Foods.

Dryer type	Usual food type
Air convection dryers	
kiln	pieces
cabinet, tray or pan	pieces, purées, liquids
tunnel	pieces

continuous conveyor belt	purées, liquids
belt trough	pieces
air lift	small pieces, granules
fluidized bed	small pieces, granules
spray	liquid, purées
Drum or roller dryers	
atmospheric	purées, liquids
vacuum	purées, liquids
Vacuum dryers	
vacuum shelf	pieces, purées, liquids
vacuum belt	purées, liquids
freeze dryers	pieces, liquids

Fruit and Vegetable Natural Drying — Sun and Solar Drying

Surplus production and specifically grown crops may be preserved by natural drying for use until the next crop can be grown and harvested. Natural dried products can also be transported cheaply for distribution to areas where there are permanent shortages of fruit and vegetables.

The methods of producing sun and solar dried fruit and vegetables described here are simple to carry out and inexpensive. They can be easily employed by grower, farmer, cooperative, etc.

The best time to preserve fruits and vegetables is when there is a surplus of the product and when it is difficult to transport fresh materials to other markets. This is especially true for crops which are very easily damaged in transport and which stay in good condition for a very short time. Preservation extends the storage (shelf) life of perishable foods so that they can be available throughout the year despite their short harvesting season.

Sun and solar drying of fruits and vegetables is a cheap method of preservation because it uses the natural resource/source of heat: sunlight. This method can be used on a commercial scale as well at the village level provided that the climate is hot, relatively dry and free of rainfall during and immediately after the normal harvesting period. The fresh crop should be of good quality and as ripe (mature) as it would need to be if it was going to be used fresh. Poor quality produce cannot be used for natural drying.

Dried fruit and vegetables have certain advantages over those preserved by other methods. They are lighter in weight than their corresponding fresh produce and, at the same time, they do not require refrigerated storage. However, if they are kept at high temperatures and have a high moisture content they will turn brown after relatively short periods of storage.

Different lots at various stages of maturity (ripeness) must NOT be mixed together; this would result in a poor dried product. Some varieties of fruit and vegetables are better for natural drying than other; they must be able to withstand natural drying without their texture becoming tough so that they are not difficult to reconstitute. Some varieties are unsuitable because they have irregular shape and there is a lot of wastage in trimming and cutting such varieties.

Damaged parts which have been attacked by insects, rodents, diseases, etc. and parts which have been discoloured or have a bad appearance or colour, must be removed. Before trimming and cutting, most fruit and vegetables must be washed in clean water. Onions are washed after they have been peeled.

Trimming includes the selection of the parts which are to be dried, cutting off and disposing of all unwanted material. After trimming, the greater part of the fruit and vegetables cut into even slices of about 3 to 7 mm thick or in halves/quarters, etc.

It is very important to have all slices/parts in one drying lot of the same thickness/size; the actual thickness will depend on the kind of material. Uneven slices or different sizes dry at different rates and this result in a poor quality end product. Onions and root crops are sliced with a hand slicer or vegetable cutter; bananas, tomatoes and other vegetables or fruit are sliced with stainless-steel knives.

As a general rule plums, grapes, figs, dates are dried as whole fruits without cutting/slicing. Some fruit and vegetables, in particular bananas, apples and potatoes, go brown very quickly when left in the air after peeling or slicing; this discolouration is due to an active enzyme called phenoloxidase. To prevent the slices from going brown they must be kept under water until drying can be started. Salt or sulphites in solution give better protection. However, whichever method is used, further processing should follow as soon as possible after cutting or slicing.

Blanching - exposing fruit and vegetable to hot or boiling water - as a pre-treatment before drying has the following advantages:

- It helps clean the material and reduce the amount of micro-organisms present on the surface;
- It preserves the natural colour in the dried products; for example, the carotenoid (orange and yellow) pigments dissolve in small intracellular oil drops during blanching and in this way they are protected from oxidative breakdown during drying;
- It shortens the soaking and/or cooking time during reconstitution.

During hot water blanching, some soluble constituents are leached out: water-soluble flavours, vitamins (vitamin C) and sugars. With potatoes this may be an advantage as leaching out of sugars makes the potatoes less prone to turning brown.

Blanching is a delicate processing step; time, temperature and the other conditions must be carefully monitored.

A suitable water-blanching method in traditional processing is as follows:

- The sliced material is placed on a square piece of clean cloth; the corners of the cloth are tied together;
- A stick is put through the tied corners of the cloth;
- The cloth is dipped into a pan containing boiling water and the stick rests across the top of the pan thus providing support for the cloth bag.

The average blanching time is 6 minutes. The start of blanching has to be timed from the moment the water starts to boil again after the cloth bag has been dipped into the pan. While the material is being blanched the cloth bag should be raised and lowered in the water so that the material is heated evenly.

When the blanching time is completed the cloth bag and its content should be dipped into cold water to prevent over-blanching. If products are over-blanched (boiled for too long) they will stick together on the drying trays and they are likely to have a poor flavour.

Green beans, carrots, okra, turnip and cabbage should always be blanched. The producer can choose whether or not potatoes need blanching. Blanching is not needed for onions, leeks, tomatoes and sweet peppers. Tomatoes are dipped into hot water for one minute when they need to be peeled but this is not blanching.

Use of Preservatives

Preservatives are used to improve the colour and keeping qualities of the final product for some fruits and vegetables. Preservatives include items such as sulphur dioxide, ascorbic acid, citric acid, salt and sugar and can either be simple or compound solutions. Treatment with preservatives takes place after blanching or, when blanching is not needed, after slicing. In traditional, simple processing the method recommended is:

- Put enough preservative solution to cover the cloth bag into a container/pan;
- Dip the bag containing the product into the preservative solution for the amount of time specified;
- Remove the bag and put it on a clean tray while the liquid drains out. The liquid which drains out must not go back into the preservative solution because it would weaken the solution.

Care must be taken after each dip to refill the container to the original level with fresh preservative solution of correct strength. After five lots of material have been dipped, the remaining solution is thrown away; *i.e.* a fresh lot of preservative solution is needed for every 5 lots of material. The composition and strength of the preservative solution vary for different fruit and vegetables.

Table. Dilutions of sodium metabisulphite with water to obtain "PM" of sulphur dioxide (SO2).

PPM SO_2	SODIUM METABISULPHITE	
	Grams per litre of water	Grams per 20 litre tin of water
1000	1.5	30.0
2000	3.0	60.0
3000	4.5	90.0
4000	6.0	120.0
5000	7.5	150.0
6000	9.0	180.0
7000	10.5	210.0

Note:

One level teaspoon of sodium metabisulphite = c. 5 g.

The strength of sulphur dioxide is expressed as "parts per million" (ppm). 1.5 grams of sodium metabisulphite in one litre of water gives 1000 ppm of sulphur dioxide. Details for solutions of different strengths are given in the following table. The preservative solutions in the fruit and vegetable pre-treatment can only be used in enamelled, plastic or stainless-steel containers; never use ordinary metal because solutions will corrode this type of container. As a general rule, preservatives are not used for treating onions, garlic, leeks, chilies and herbs.

Osmotic Dehydration

In osmotic dehydration the prepared fresh material is soaked in a heavy (thick liquid sugar solution) and/or a strong salt solution and then the material is sun or solar dried. During osmotic treatment the material loses some of its moisture. The syrup or salt solution has a protective effect on colour, flavour and texture. This protective effect remains throughout the drying process and makes it possible to produce dried products of high quality. This process makes little use of sulphur dioxide.

Sun Drying

The main problems for sun drying are dust, rain and cloudy weather. Therefore, drying areas should be dust-free and whenever there is a threat of a dust storm or rain, the drying trays should be stacked together and placed under cover.

In order to produce dust-free and hygienically clean products, fruit and vegetable material should be dried well above ground level so that they are not contaminated by dust, insects, livestock or people. All materials should be dried on trays designed for the purpose; the most common drying trays have wooden frames with a fitted base of nylon mosquito netting. Mesh made of woven grass can also be used. Metal netting must NOT be used because it discolours the product.

The trays should be placed on a framework at table height from the ground. This allows the air to circulate freely around the drying material and it also keeps the food product well away from dirt. Ideally the area should be exposed to wind and this speed up drying, but this can only be done if the wind is free of dust.

With 80 cm x 50 cm trays, the approximate load for a tray is 3 kg; the material should be spread in even layers. During the first part of the drying period, the material should be stirred and turned over at least once an hour.

This will help the material dry faster and more evenly, prevent it sticking together and improve the quality of the finished product. Products for sun drying should be prepared early in the day; this will ensure that the material enjoys the full effect of the sun during the early stages of drying.

At night the trays should be stacked in a ventilated room or covered with canvas. Plastic sheets should NEVER be used for covering individual trays during sun drying. Dry or nearly dry products can be blown out of the tray by the wind. However, this can be protected by covering the loaded tray with an empty one; this also gives protection against insects and birds.

Shade Drying

Shade drying is carried out for products which can lose their colour and/or turn brown if put in direct sunlight. Products which have naturally vivid colours like herbs, green and red sweet peppers, chilies, green beans and okra give a more attractive end-product when they are dried in the shade. The principles for the shade drying are the same as for sun drying. The material to be dried requires full air circulation. Therefore, shade drying is carried out under a roof or thatch which has open sides; it cannot be done either inside conventional buildings with side walls or in compounds sheltered from wind. Under dry conditions when there is a good circulation of air, shade drying takes little more time than is normally required for drying in full sunlight.

Identification of Suitable Designs of Solar Dryers for Different Applications

In the selection of appropriate solar dryers for commercial scale operation, it is imperative that economics be kept in view at all time.

A total Energy System concept should be employed and due consideration be given to parasitic energy consumption. The following features have been identified:

- Large scale dryers are more promising than small scale ones. However, small scale dryers should not be neglected.
- The dryer should be designed to maximise the utilisation factor of the capital investment, *i.e.* multi-products (fruit, vegetables and other raw material) and multi-use (*e.g.* drying and heating water for domestic use).
- In general, an auxiliary heat source should be provided to assure reliability, to handle peak loads and also to provide continuous drying during periods of no sunshine.
- Forced convection indirect dryers are preferred because they offer better control, more uniform drying and because of their high heat collection efficiency result in smaller collector area. However, parasitic power should be kept to a minimum.

Two dryer systems have been identified:

- A cabinet type dryer with natural convection for internal air circulation for the processing of dried fruit such as mango, banana, pineapple, apricot, pear, apple, etc. and also for potato chips and other vegetables;
- A greenhouse type dryer with forced air circulation.

Some of the barriers to the commercial development of solar dryers have been attributed to:

- Initial cost - poor farmers cannot afford them
- Durability - constant breakdown due to using low cost building materials
- Misuse - through lack of training and technical skills
- Dependability and reliability - during the wet season when drying is critical there is not enough solar energy available
- The wider use of Solar Drying Systems has been limited by other factors which are not necessarily of a technical or technological nature. Among the most important are the lack of national policies directed to promoting the drying of produce at the production site, in order to reduce losses, improve quality and increase farmers' earnings.

Construction of Solar Dryers

In the case of simple natural convection dryers it may be more appropriate to build and operate a number of small units. Multiplicity allows diversity, since more than one crop can be dried at a time. A further advantage is that if one dryer is out of operation due to damage, drying can still continue at reduced capacity using the other dryers. On the other hand, more sophisticated dryers, such as forced convection solar dryers, benefit from economies of scale due to the investment tied up in the fan and the source of heat.

Generally speaking, one large dryer will be more cost-effective than two smaller units. However, it should be taken into consideration that an oversized unit will be operating at less than full capacity, reducing any cost advantage. The drying area required will depend on local conditions, commodity and number of trays on each rack or trolley.

Construction Methods and Materials

Construction methods and available materials may vary considerably from location to location. It is not within the scope of this document to discuss individual, local circumstances. Some general guidelines regarding factors which must be considered can, however, be given:

- *Dimensions of standard materials:* Where possible, design should take account of the sizes of material locally available. For example, it would be poor design to specify the width of a corrugated iron collector as 1.1 m if the standard width of a corrugated iron sheet is 1 m.

Before finalising a design the commercial availability of materials must be ascertained.

- *Use of rural materials:* The cost of building of solar dryer can be minimised if the producer is able to use wood cut straight from the forest rather than prepared timber.

Careful design in the development stage of a dryer can often facilitate the use of cheaper materials. Difficulties caused by these materials are in joining pieces of the structure, in sealing the structure against air leaks, and in attaching the plastic sheet to the (wooden) frame. There is obvious scope for designs which use prepared timber for strategic points and unprepared at others. Where the use of wood is necessary, remember to take environmental factors into consideration. For example, determine the effect of flash flooding or termites might be and take the appropriate preventive action.

- *Use of plastic sheets:* For many solar dryers, the clear plastic sheet used is the major capital cost to the farmer; therefore, the type of plastic chosen is important.

A choice must be made between a relatively cheap plastic such as ordinary polyethylene which will last, at best, for one season due to photo-degradation and wear and tear; and a more expensive, better quality plastic less prone to photo-degradation; or even glass or a rigid plastic.

Attaching plastic sheet to the framework structure, so as to minimise the likelihood of the plastic being torn is, perhaps, the most difficult part of building a dryer. Listed below are some general points which should be followed to prolong the useful lifetime of plastic sheet on a solar dryer:

- When attaching plastic sheet to the framework, care should be taken to stretch the plastic at the points of attachment, but the plastic should not be so loose that it will flap about in the wind;
- Rather than merely stapling or nailing the plastic directly to the framework, it is preferable to sandwich the plastic between the framework and a batten. This may not be practical when unprepared wood or other materials are being used.
- No sharp edges should come in contact with the plastic sheet since these will initiate tears;
- Fold over the plastic at the point of attachment to the frame, so that there are two or more layers of plastic. This will help prevent tears;
- When fixing the sheet over the framework, sags and hollows in which water can collect should be avoided wherever possible;
- The dryer should be handled as carefully and as seldom as possible during operation and when not in use.

Technical Criteria

The following design factors must be established:

- The throughput of the dryer over the productive season; the size of batch to be dried;
- The drying period(s) under stated conditions;
- The initial and desired final moisture content of the commodity (if known);
- The drying characteristics of the commodity, such as maximum drying temperature, effect of sunlight upon the product quality, etc.;
- Climatic conditions during the drying season, *i.e.* sunlight intensity and duration; air temperature and humidity; wind speed (such data may be available from local meteorological stations);
- Availability and reliability of electrical power;

- The availability, quality, durability and price of potential construction materials such as:
  - Glazing materials: glass, plastic sheet or film;
  - Wood (prepared or unprepared);
  - Nails, screw, bolts, etc.;
  - Metal sheet, flat or corrugated angle iron;
  - Bricks (burnt or mud), concrete blocks, stones, cement, sand, etc.
  - Roofing thatch;
  - Metal mesh, wire netting, etc.
  - Mosquito netting, muslin, etc.
  - Bamboo or fibre weave;
  - Black paint, other blackening materials;
  - Insulation material; sawdust, etc.;
- The type of labour available to build and operate the dryer;
- The availability of clean water at the site for preparation of the commodity prior to drying.

In any one situation there may well be other technical factors that need to be considered.

Socio-economic Criteria

From the initial considerations, estimates of the capital costs of the dryer, the price of the commodity to be dried, and the likely selling price of the dried product will have been made. Other question that need to be considered are the following:

- Who will own the dryer?
- Is the dryer to be constructed by the end-user (with or without advice from extension agencies), local contractors, or other organisations?
- Who will operate and maintain it?
- How can the drying operation be incorporated into current practices?
- Are sources of finance from local authorities or extension agencies available, etc.?

Obviously there are many other socio-economic factors, particularly those of a local nature, which must be taken into account.

It cannot be stressed too highly that if such factors are not taken into account and evaluated, then is every chance that an inappropriate dryer design may result. Equal emphasis must be placed on both technical and socio-economic factors.

- Situations where solar dryer may be useful:
  - Where the cost of conventional energy is prohibitive and/or the supply is erratic, to supplement existing artificial drying systems and reduce fuel costs;
  - Where land is in short supply or expensive;
  - Where the quality of existing sun dried products can be improved upon;
  - Where the labour is in short supply;
  - Where is plenty of sunshine, but high humidity.
- Situations where solar dryers may not be useful:
  - Where conventional energy sources are abundant and cheap;
  - Where large amounts of combustible by-products or waste materials are freely available;
  - Where there is insufficient sunshine;
  - Where is plenty of sunshine and arid conditions (sun drying may suffice);
  - Where the quality of sun dried products already made cannot be improved upon;
  - Where local operators are insufficiently trained;
  - Where the ramifications of introducing a solar dryer have not been completely thought out.

Sun/solar Drying Tray

The drying tray described requires seasoned timber 22.5 mm thick x 50 mm wide. A sun drying tray requires 6 metres of seasoned timber 22.5 mm thick x 50 mm wide. Cut the timber into lengths of 900 mm long for the sides of the tray and 600 mm long for the ends - 4 pieces of each length will be needed. The ends of each piece are cut as shown in the drawing - this is to make flush fitting joints. Join the corners using small brass screws 20 mm long.

To make extra strong joints use good quality wood glue as well as the screws. The nylon mosquito netting or grass woven mesh can be fitted between the frames as shown in the bottom drawing. Cut the mesh a little larger than the size of the frame. Using drawing pins, pin the mesh to the OUTSIDE edges of one of the frames - the mesh should be pulled tight as the pins are put in around the edges. Lay the other frame on top and drill holes about 3 mm in diametre at the points marked X in the top drawing. Use nails that are a tight fit in the holes and tap gently into place leaving a portion standing above the frame. Cut off the standing part to leave a piece about 12 mm long which is then bent over and tapped firmly down onto the frame. When the frame has been put together tightly, the drawing pins can be removed.

6

Investigation of Foodborne Disease Outbreaks

Infections vs. Intoxications

- Infections characterized by
  - Viable cells consumed
  - Relatively long onset time for symptoms
  - Relatively long duration
  - Fever a common symptom (body's response)
- Intoxications characterized by
  - Viable cells may not be present
  - Quicker onset time (if preformed toxin)
  - Often short duration (unless toxin produced in body)
  - No fever (usually)
- But
  - Invaders make factors that help them adhere and penetrate
  - Toxins produced by many pathogens contribute to symptoms
  - Many organisms produce many different extracellular products (enzymes, toxins)
- So
  - Intoxications are more common than you think!

Toxins

- Terms
  - Exotoxin = extracellular protein toxin
  - Endotoxin = lipid A portion of Gram-neg outer memb.
  - Enterotoxin = toxin that acts on gastrointestinal tract, producing typical food poisoning symptoms
- Nomenclature
  - Named for host cell attacked: cytotoxin, neurotoxin
  - Named for producer or disease: cholera, Shiga
  - Named for activity: lecithinase, adenylate cyclase

- Letter designation: exotoxin A

- Types
  - A-B toxins
    a. Reduce temperature → slow free radical
    b. B portion binds to receptor => specificity
    c. A portion enters cells, has enzymatic activity
    d. Examples:
    e. Cholera toxin
    f. *E. coli* LT (heat-labile toxin)
  - Membrane-disrupting toxins
    a. Nonenzymatic
    b. Insert into membrane to form pores
    c. Enzymatic attack on membrane components
    d. Example: phospholipase (hemolysin)
  - Superantigens
    a. Cause hyperimmune response by the host
    b. Example: toxic shock toxin of S. aureus

Role of Toxins in Foodborne Disease

- Consumed as preformed => self-limiting
- Produced by colonized bacteria => local or distal effect
- Produced by infecting bacteria to aid invasion
- Autoimmune response to superantigens
- Endotoxin elicits immune response => shock

How to Show an Organism Causes an intoxication

- Feed culture and produce symptoms
- Feed culture filtrate and produce symptoms

Organisms Recognized for Foodborne Intoxications

- *Staphylococcus aureus* - enterotoxins
- *Clostridium botulinum* - neurotoxin
- *Clostridium perfringens* - enterotoxin, other toxins
- *Bacillus cereus* - diarrheagenic and emetic toxins
- *Vibrio cholerae* - cholera toxin
- enterotoxigenic *E. coli* - heat-stable and heat-labile toxins

- *Aspergillus flavus, A. parasiticus* - aflatoxins
- Fusarium sp. - fumonisin, zearalenone
- Aspergillus, Penicillium sp. - ochratoxins, patulin
- Alternaria - various toxins
- *Gonyaulax catenella, G. tamarensis* - saxitoxin

Agents of Foodborne Infections that also Produce Toxins

- Shigella sp. - shiga toxin
- *Listeria monocytogenes* - listeriolysin
- Salmonella sp. - enterotoxin, cytotoxin
- Enterohemorrhagic E. coli - shiga-like toxin
- *Vibrio parahaemolyticus* - hemolysin
- *Yersinia enterocolitica* - enterotoxin
- *Campylobacter jejuni* - enterotoxin
- *Aeromonas hydrophila* - multifunction toxin
- *Plesiomonas shigelloides* - enterotoxin

Bacterial Agents of Food Born Illnesses

The Organism

- Gram-positive cocci
- Facultative
- Grows 10-45 °C (opt. 40-45)
- Grows pH 4.5 - 9.3 (opt. 6-7)
- Grows to aw 0.83
- Requires organic growth factors
- Salt-tolerant
- Heat-sensitive
- Doesn't compete well with normal flora of foods

Diseases

- Skin abscesses
- Scalded skin syndrome
- Toxic shock syndrome
- Nosocomial infections
- Food poisoning
  - Vomiting, diarrhea, nausea, cramps

- Onset 1-6 hr
- Recovery 1-3 days

Incidence in Environment and Foods

- Carried by animals and humans - skin, nasal cavity
- Foods of animal origin
- Foods receiving significant handling
- Frequently associated with outbreaks:
  - Baked ham
  - Pork
  - Salads
  - Pastries
  - Dried milk
  - Whey
- Used as indicator of proper handling, processing
- Much work done on detection and enumeration methods
  - Causes of sublethal injury
  - Methods to recover injured cells
- Risk factor: improper storage temperature
- Prevention: don't let organisms multiply in food

Extracellular Enzymes

- > 34 different extracellular proteins
- Hydrolytic enzymes: proteases, lipases, hemolysins
- Penicillinase
- Coagulase
  - Clot formation in serum
  - Common identification test for *S. aureus*
  - Correlation with enterotoxin? < 100%
- Heat-stable nuclease
  - Attack RNA and DNA
  - Used to identify *S. aureus*
  - Correlation with enterotoxin? < 100%
- Toxins
  - Enterotoxins
  - Toxic shock toxin

Enterotoxins

- Merlin Bergdoll, Food Research Institute, Univ. Wisconsin
- Types: A, B, C1, C2, C3, D, E (SEA, SEB, etc.)
  - SEA most commonly found in outbreaks
  - SEB produced in high (but variable) amounts
- Structure
  - Single polypeptide chain
  - MW 26,360 - 28,500 (230 -239 amino acids)
- Mode of action
  - Stimulates vagus nerves in intestine?
  - Superantigen effects?
- Resistant to heat, radiation, proteolytic enzymes
- Toxic dose ~ 1 5g
- Production
  - Favored by optimum growth conditions
  - Produced during late log and stationary phases
  - Need about 107 cells/g to get toxic dose

Toxin Detection

- Traditional animal assays
  - Monkey feeding
  - Injection of kittens i.p. or i.v.
  - Expensive, time-consuming, unpleasant
  - Sensitive, only measure of biological activity
- Immunological assays
  - Precipitation reactions (Ouchterlony, microslide)
  - Hemagglutination
  - ELISA, RIA
  - Problems with recovery from food and sensitivity
  - Must be able to recover and detect 1 ng/g food

Clostridium botulinum

The Organism

- Gram-positive, anaerobic, spore-forming rods
- Defined and grouped by toxin production: A - G

- Grouped by physiology
  - Group I: proteolytic, high heat resistance
  - Group II: nonproteolytic, low heat resist., grow in cold
  - Group III: nonproteolytic
  - Group IV: little known

Occurrence in Environment and in Food

- Soil and sediment
- Fish
- Meats
- Fruits, vegetables, mushrooms

Growth Characteristics

- Temp. range: 10-50 C (aquatic strains grow to 3 C)
- Minimum aw: 0.94 (0.97 for aquatic strains)
- Minimum pH: 4.7-4.8
- Salt tolerance: up to 10% NaCl (6% for aquatic strains)
- Strict anaerobes
- Complex nutritional requirements
- Heat-resistant spores (especially type A)
- Don't compete well with normal flora
- Favored when heat has killed competitors

Potentially Dangerous Foods

- Heat treatment that kills vegetative cells but not spores
- Restricted oxygen (vacuum-packaged, modified atm)
- Low-acid
- Nutrient-rich

The Illness

- Categories
  - Foodborne
  - Infant botulism
  - Wound botulism
  - Unclassified
- Syndrome

  - Onset in 12-36 hr (avg)
  - Fatigue
  - Nausea and vomiting (early)
  - Poor coordination
  - Neurological signs
  - Respiratory failure
- Treatment
  - Try to clear toxin from GI tract
  - Antitoxin
  - Support therapy
- Prevention
  - Kill spores or prevent outgrowth
  - Specific processes for different food categories

The Toxin

- Structure
  - Progenitor vs. derivative forms (S, M, L, LL forms)
  - Monomer = 150,000 MW
  - Heavy chain: receptor and transfer functions
  - Light chain: endopeptidase activity
- Mode of action
  - Binds to and enters neurons at synapse
  - Interferes with release of neurotransmitters
- Genetic control
  - C1 and D carried by phage => phage conversion
  - G may be plasmid-borne
  - others not located yet

Other Toxins

- C2 - may increase permeability of cells to toxin
- C3 - may affect some cell regulatory mechanisms

Lab Testing

- Look for presence of toxin
  - Directly in food

  - In cultures of organisms grown from food
- Mouse lethality test
  - Inject intraperitoneally
  - Trypsin treatment
  - Add antisera vs. A, B, E
  - Look for death in < 4 days
- ELISA, other immunological assays in development
  - Not as sensitive
  - Problems in toxin extraction
  - Don't detect activity

Clostridium Perfringens

The organism

- Characteristics
  - Gram-positive
  - Sporeforming rod
  - Encapsulated
  - Anaerobic
- Types: A - E, based on toxin production
- Temperature effects
  - Growth range 15-50C (opt 43-45C)
  - High temperature:
    a. Vegetative cells inactivated >60C
    b. Low- vs. high-heat resistant spores
    c. Heat-activation for spore germination
  - Low temperature:
    a. Cold-sensitive vegetative cells
    b. Lose viability in stored samples
- O/R potential
  - Need anaerobic conditions for growth
  - Survive air exposure
- Aw: minimum ~0.95
- pH: optimum 6-7; range 5-8
- Salts: combined inhibition of NaCl + curing salts

Illnesses

- Foodborne disease
  - 8-24 hr onset
  - 24 hr duration
  - Diarrhea + abdominal pain
- Pig-bel
  - Traditional pig feasts in Papua New Guinea
  - Type C (from pig) causative agent
  - 'Toxin (necrotizing) => bloody diarrhea
- Gas gangrene
  - Infection of deep wounds (anaerobic)
  - Tissue necrosis + gas production
- Antibiotic-associated diarrhea
  - Normal flora reduced
  - Resident C. *perfringens* proliferate

Foodborne Illness

- Reservoirs for the organism
  - Soil & dust
  - Intestine
  - Foods: meats, spices
- Outbreaks
  - Variable reporting
  - Frequently large (food service source)
  - Meat & poultry most common vehicles
  - Improper cooking, cooling
  - Prevention: temperature control
- History
  - Association with gas gangrene known since 1900
  - Food poisoning role suggested in 1940s
  - Study of organism in 1950s
  - Human feeding studies
    a. Remove off-flavors
    b. 1953: live cultures fed effective dose = 4-6 x 109 veg. cells

c. 1971: ileal loop activity vs. human diarrhea

d. 1977: 8-10 mg toxin induce symptoms neutralize stomach with bicarbonate

Pathogenicity

- Mode of action - macroscopic level
  - Fluid accumulation in intestine => diarrhea
  - Increased net secretion of fluid, Na+, Cl-
  - Glucose absorption inhibited
  - Damage to vili tips, brush borders
- Specific site of action
  - Disruption of plasma membrane
  - Toxin binds to membrane via 50,000 MW protein
  - Functional holes made
  - Altered membrane permeability
- Toxin
  - Protein, 34,000 MW, 309 amino acids
  - Activated by trypsin: 4000 MW peptide removed
  - Released during sporulation
- Assays
  - Ileal loop
  - Animal diarrhea
  - Skin permeability
  - Serological

Bacillus Cereus

Theorganism

- Characteristics
  - Gram-positive
  - Sporeformer
  - Facultative
  - Very large cells
  - Related to:
- B. mycoides B. anthracis B. thuringiensis
  - Temp range 5-50C, opt. 35-40C

- Short generation time (20-30 min)
- pH range 4.5-9.3
- Sporulates and germinates easily
- High heat and drying resistance of spores
- Acid from sugar fermentations (except mannitol)
- Mildly proteolytic

- Occurrence
  - Wide distribution in soil, dust, air
  - Carried by humans and animals (from foods?)
  - In many food products:
    a. Dairy products
    b. Rice and cooked oriental foods
    c. Spices and spice mixes
    d. Dried products (flour, dry milk, pudding, soup mix)
    e. Beans and bean sprouts
    f. Meats
    g. Bakery products (cream-filled pastries)
- Extracellular products
  - Proteases (sweet curdling of dairy products)
  - '-lactamases (penicillin resistance)
  - Phospholipases
  - Hemolysins
  - Toxins
    a. Lethal toxin: kills mice injected i.v.
    b. Enterotoxins: diarrheal and emetic

Pathogenesis

- Infections in humans and animals
  - Wound infections
  - Septicemia
  - Mastitis
  - Meningitis
- Diarrheal food poisoning
  - Early reports in Europe in early 1900's
  - Linked to foodborne outbreaks in 1950's

- Hauge (Norway)
    a. Lethal toxin: kills mice injected i.v.
- First complete account of outbreaks
    a. Drank culture in vanilla sauce -> symptoms
    b. Recognized worldwide by 1970's
- U.S. outbreaks: 35 reported for 1977-1984
    a. Recognized worldwide by 1970's
    b. Meat loaf cooked rice
    c. Mashed potatoes green bean salad
    d. Chicken pot pie vanilla sauce
    e. Vegetable sprouts turkey loaf
- High numbers needed (> 105/g)
- Syndrome
    a. Vegetable sprouts turkey loaf
    b. Watery diarrhea, cramps, abdominal pain
    c. Nausea but not vomiting, no fever
    d. Onset 12-18 hr
    e. Duration <24 hr
    f. => similar to C. perfringens
- Emetic food poisoning
    a. First outbreak in 1971 (England, Chinese restaurant)
    b. 110 outbreaks by 1979, most in Chinese restaurants or carry-outs
    c. Implicated food: fried rice (boiled rice kept at r.t.)
    d. Syndrome
    e. Acute nausea and vomiting
    f. Diarrhea not common
    g. No fever
    h. Onset 1-5 hr
    i. Duration 6-24 hr
    j. => similar to S. aureus
    k. BUT B. cereus (high counts), not Staph, found in foods

Surveillance

- Toxins: no reliable assays available for routine use

- Organism
  - Ubiquitous => mere presence means little
  - High counts can indicate potential problems
  - How many isolates make toxins?
    a. Probably most make diarrheal toxin
    b. Unknown frequency for emetic toxin
- Methods
  - May need enrichment for routine screening
  - Selective/differential medium based on
    a. Resistance to polymyxin
    b. Inability to ferment mannitol
    c. Lecithinase (phospholipase) production

Prevention

- Can't keep the organism out of foods
- Not dangerous at low levels => focus on control of growth
- Processing/cooking should
  - Destroy vegetative cells
  - Prevent spore germination and outgrowth
  - Prevent further multiplication
  - Thorough cooking should kill cells and spores
- Slow cooking, temps at 100C can allow spore survival
- Avoid nonrefrigerated storage of rice and other cereals
- Prepare food in small quantities for rapid heating, cooling

Escherichia Coli

Diarrheagenic Types

- EPEC (enteropathogenic)
  - First pathogenic type identified
  - Severe illness in infants
- ETEC (enterotoxigenic)
  - Similarities to cholera
- EIEC (enteroinvasive)
  - Similarities to Shigella
- EHEC (enterohemorrhagic)

    - O157:H7
- EAggEC (enteroaggregative)
    - Newest identified type

ETEC

- History
    - Identified in 1950s: similar to cholera in ileal loops
    - 1960s and 1970s:
        a. Toxins identified (LT and ST)
- Epidemiology and prevalence travelers diarrhea, infant diarrhea diarrhea in domestic animals, esp. piglets
- Food and water transmission
    - Large inoculum (108) needed for illness
    - Usual vehicles: raw vegetables, ice, water
- Disease
    - Noninflammatory diarrhea, nausea, cramping
    - Mild (1-5 days) to severe (cholera-like)
    - Endemic in developing world - childhood diarrhea
    - Immunity in older children and adults - carriers?
    - Not common in industrialized countries
- Virulence factors
    - Colonization factors (CFAs)
        a. Needed to adhere to proximal small intestine
        b. Fimbriae mediate attachment
        c. Species specificity
    - Toxins
        a. Heat-labile toxin (LT)
        b. Heat-stable toxin (ST)

Cholera Toxin (a digression)

- Mode of action
    - Colonization in the small intestine, toxin production
    - Normal intestinal function
        a. Fluid absorption/secretion controlled by ion conc.
        b. Specific pumps for Na+, Cl-, K+, HCO3-
        c. Net flow for ions + water is lumen -> tissues

  - Add organisms or cholera toxin
    a. No membrane damage
    b. Na+ absorption reduced,
    c. Cl- secretion increased
    d. Net water flow into lumen
- Structure
  - A-B type: 5 B-subunits + 1 A-subunit
  - nick A to form A1 + A2 -> release A1 (active) part
- Host cell interaction
  - B-subunits bind to cell surface receptor (ganglioside)
  - A1 must enter cell and reach its target
- A1 activity
  - ADP-ribosylation of GS membrane protein
  - cAMP production continues uncontrolled
  - Various transport processes affected by cAMP
- Genetics and control
  - Structural genes (ctxA, ctxB) in operon
  - May have multiple copies -> more toxin produced

Enterotoxins of ETEC

- LT (heat-labile toxin)
  - Inactivated at 60C for 30 min
  - 5 B + 1 A subunit C LT-I (humans, pigs) - very similar to cholera toxin
  - LT-II (water buffalo) - different protein
  - Same activity as cholera toxin
  - Usually plasmid-borne
- ST (heat-stable toxin)
  - Active after boiling for 30 min
  - Small molecule (~2 kDa), internal cross-links
  - STa (human strains, methanol-soluble)
  - STb (porcine strains, methanol-insoluble)
  - Nonantigenic, not affected by anti-LT or anti-CT
  - Stimulates guanylate cyclase => accum. cGMP
- Detection

- Animal assays
  ileal loop for LT
  suckling mouse for ST
- other assays for LT
  tissue culture - morphological changes
  immunoassays
  DNA probes

EHEC and Shiga-like toxin

- Shiga toxin produced by S. dysenteriae
  - A-B type: 5 B subunits + 1 A subunit
  - A subunit affects ribosome -> protein synthesis stops
  - has enterotoxic, neurotoxic, cytotoxic activities, too

Salmonella spp

1. *Name of the Organism: Salmonella* spp. Salmonella is a rod-shaped, motile bacterium — nonmotile exceptions *S. gallinarum* and *S. pullorum*—, nonsporeforming and Gram-negative. There is a widespread occurrence in animals, especially in poultry and swine. Environmental sources of the organism include water, soil, insects, factory surfaces, kitchen surfaces, animal feces, raw meats, raw poultry, and raw seafoods, to name only a few.
2. *Nature of Acute Disease: S. typhi* and the paratyphoid bacteria are normally caused septicemic and produce typhoid or typhoid-like fever in humans. Other forms of salmonellosis generally produce milder symptoms.
3. *Nature of Disease:* Acute symptoms — Nausea, vomiting, abdominal cramps, minal diarrhea, fever, and headache. Chronic consequences — arthritic symptoms may follow 3-4 weeks after onset of acute symptoms. Onset time — 6-48 hours. Infective dose — As few as 15-20 cells; depends upon age and health of host, and strain differences among the members of the genus. Duration of symptoms — Acute symptoms may last for 1 to 2 days or may be prolonged, again depending on host factors, ingested dose, and strain characteristics. Cause of disease — Penetration and passage of Salmonella organisms from gut lumen into epithelium of small intestine where inflammation occurs; there is evidence that an enterotoxin may be produced, perhaps within the enterocyte.
4. *Diagnosis of Human Illness:* Serological identifica-tion of culture isolated from stool.

5. *Associated Foods:* Raw meats, poultry, eggs, milk and dairy products, fish, shrimp, frog legs, yeast, coconut, sauces and salad dressing, cake mixes, cream-filled desserts and toppings, dried gelatin, peanut butter, cocoa, and chocolate. Various Salmonella species have long been isolated from the outside of egg shells. The present situation with *S. enteritidis* is complicated by the presence of the organism inside the egg, in the yolk. This and other information strongly suggest vertical transmission, i.e., deposition of the organism in the yolk by an infected layer hen prior to shell deposition. Foods other than eggs have also caused outbreaks of *S. enteritidis* disease.

   It is estimated that from 2 to 4 million cases of salmonellosis occur in the U.S. annually.

   The incidence of salmonellosis appears to be rising both in the U.S. and in other industrialized nations. *S. enteritidis* isolations from humans have shown a dramatic rise in the past decade, particularly in the northeast United States (6-fold or more), and the increase in human infections is spreading south and west, with sporadic outbreaks in other regions.
6. *Relative Frequency of Disease:* Reported cases of Salmonellosis in the U.S. excluding typhoid fever for the years 1988 to 1995. The number of cases for each year varies between 40,000 and 50,000.
7. *Complications: S. typhi* and *S. paratyphi* A, B, and C produce typhoid and typhoid-like fever in humans. Various organs may be infected, leading to lesions. The fatality rate of typhoid fever is 10% compared to less than 1% for most forms of salmonellosis. *S. dublin* has a 15% mortality rate when septicemic in the elderly, and *S. enteritidis* is demonstrating approximately a 3.6% mortality rate in hospital/ nursing home outbreaks, with the elderly being particularly affected. Salmonella septicemia has been associated with subsequent infection of virtually every organ system. Postenteritis reactive arthritis and Reiter's syndrome have also been reported to occur generally after 3 weeks. Reactive arthritis may occur with a frequency of about 2% of culture-proven cases. Septic arthritis, subsequent or coincident with septicemia, also occurs and can be difficult to treat.
8. *Target Populations:* All age groups are susceptible, but symptoms are most severe in the elderly, infants, and the infirm. AIDS patients suffer salmonellosis frequently (estimated 20-fold more than general population) and suffer from recurrent episodes.
9. *Foods Analysis:* Methods have been developed for many foods having prior history of Salmonella contamination. Although

conventional culture methods require 5 days for presumptive results, several rapid methods are available which require only 2 days.

10. *Selected Outbreaks:* In 1985, a salmonellosis outbreak involving 16,000 confirmed cases in 6 states was caused by low fat and whole milk from one Chicago dairy. This was the largest outbreak of foodborne salmonellosis in the U.S. FDA inspectors discovered that the pasteurization equipment had been modified to facilitate the running off of raw milk, resulting in the pasteurized milk being contaminated with raw milk under certain conditions. The dairy has subsequently disconnected the cross-linking line. Persons on antibiotic therapy were more apt to be affected in this outbreak. In August and September, 1985, *S. enteritidis* was isolated from employees and patrons of three restaurants of a chain in Maryland. The outbreak in one restaurant had at least 71 illnesses resulting in 17 hospitalizations. Scrambled eggs from a breakfast bar were epidemiologically implicated in this outbreak and in possibly one other of the three restaurants. The plasmid profiles of isolates from patients all three restaurants matched. The Centers for Disease Control (CDC) has recorded more than 120 outbreaks of *S. enteritidis* to date, many occurring in restaurants, and some in nursing homes, hospitals and prisons. In 1984, 186 cases of salmonellosis (*S. enteritidis*) were reported on 29 flights to the United States on a single international airline. An estimated 2,747 passengers were affected overall. No specific food item was implicated, but food ordered from the first class menu was strongly associated with disease.

Campylobacter JejunI

1. *Name of the Organism: Campylobacter jejuni* (formerly known as *Campylobacter fetus* subsp. jejuni) *Campylobacter jejuni* is a Gram-negative slender, curved, and motile rod. It is a microaerophilic organism, which means it has a requirement for reduced levels of oxygen. It is relatively fragile, and sensitive to environmental stresses (e.g., 21% oxygen, drying, heating, disinfectants, acidic conditions). Because of its microaerophilic characteristics the organism requires 3 to 5% oxygen and 2 to 10% carbon dioxide for optimal growth conditions. This bacterium is now recognized as an important enteric pathogen. Before 1972, when methods were developed for its isolation from feces, it was believed to be primarily an animal pathogen causing abortion and enteritis in sheep and cattle. Surveys have shown that *C. jejuni* is

the leading cause of bacterial diarrheal illness in the United States. It causes more disease than *Shigella* spp. and *Salmonella* spp. combined. Although *C. jejuni* is not carried by healthy individuals in the United States or Europe, it is often isolated from healthy cattle, chickens, birds and even flies. It is sometimes present in non-chlorinated water sources such as streams and ponds. Because the pathogenic mechanisms of *C. jejuni* are still being studied, it is difficult to differentiate pathogenic from nonpathogenic strains. However, it appears that many of the chicken isolates are pathogens.

2. *Name of Disease:* Campylobacteriosis is the name of the illness caused by *C. jejuni*. It is also often known as campylobacter enteritis or gastroenteritis.
3. *Major Symptoms: C. jejuni* infection causes diarrhea, which may be watery or sticky and can contain blood (usually occult) and fecal leukocytes (white cells). Other symptoms often present are fever, abdominal pain, nausea, headache and muscle pain. The illness usually occurs 2-5 days after ingestion of the contaminated food or water. Illness generally lasts 7-10 days, but relapses are not uncommon (about 25% of cases). Most infections are self-limiting and are not treated with antibiotics. However, treatment with erythromycin does reduce the length of time that infected individuals shed the bacteria in their feces. The infective dose of *C. jejuni* is considered to be small. Human feeding studies suggest that about 400-500 bacteria may cause illness in some individuals, while in others, greater numbers are required. A conducted volunteer human feeding study suggests that host susceptibility also dictates infectious dose to some degree. The pathogenic mechanisms of *C. jejuni* are still not completely understood, but it does produce a heat-labile toxin that may cause diarrhea. *C. jejuni* may also be an invasive organism.
4. *Isolation Procedures: C. jejuni* is usually present in high numbers in the diarrheal stools of individuals, but isolation requires special antibiotic-containing media and a special microaerophilic atmosphere (5% oxygen). However, most clinical laboratories are equipped to isolate Campylobacter spp. if requested.
5. *Associated Foods: C. jejuni* frequently contaminates raw chicken. Surveys show that 20 to 100% of retail chickens are contaminated. This is not overly surprising since many healthy chickens carry these bacteria in their intestinal tracts. Raw milk is also a source of infections. The bacteria are often carried by healthy cattle and by flies on farms. Non-chlorinated water may also be a source of

infections. However, properly cooking chicken, pasteurizing milk, and chlorinating drinking water will kill the bacteria.

6. *Frequency of the Disease: C. jejuni* is the leading cause of bacterial diarrhea in the U.S. There are probably numbers of cases in excess of the estimated cases of salmonellosis (2- to 4,000,000/year).
7. Complications: Complications are relatively rare, but infections have been associated with reactive arthritis, hemolytic uremic syndrome, and following septicemia, infections of nearly any organ. The estimated case/fatality ratio for all *C. jejuni* infections is 0.1, meaning one death per 1,000 cases. Fatalities are rare in healthy individuals and usually occur in cancer patients or in the otherwise debilitated. Only 20 reported cases of septic abortion induced by *C. jejuni* have been recorded in the literature. Meningitis, recurrent colitis, acute cholecystitis and Guillain-Barre syndrome are very rare complications.
8. *Target Populations:* Although anyone can have a *C. jejuni* infection, children under 5 years and young adults (15-29) are more frequently afflicted than other age groups. Reactive arthritis, a rare complication of these infections, is strongly associated with people who have the human lymphocyte antigen B27 (HLA-B27).
9. *Recovery from Foods:* Isolation of *C. jejuni* from food is difficult because the bacteria are usually present in very low numbers (unlike the case of diarrheal stools in which 10/6 bacteria/gram is not unusual). The methods require an enrichment broth containing antibiotics, special antibiotic-containing plates and a microaerophilic atmosphere generally a microaerophilic atmosphere with 5% oxygen and an elevated concentration of carbon dioxide (10%). Isolation can take several days to a week.
10. *Selected Outbreaks:* Usually outbreaks are small (less than 50 people), but in Bennington, VT a large outbreak involving about 2,000 people occurred while the town was temporarily using an non-chlorinated water source as a water supply. Several small outbreaks have been reported among children who were taken on a class trip to a dairy and given raw milk to drink. An outbreak was also associated with consumption of raw clams. However, a survey showed that about 50% of infections are associated with either eating inadequately cooked or recontaminated chicken meat or handling chickens. It is the leading bacterial cause of sporadic (non-clustered cases) diarrheal disease in the U.S. In April, 1986, an elementary school child was cultured for bacterial pathogens (due to bloody diarrhea), and *C. jejuni* was isolated. Food

consumption/gastrointestinal illness questionnaires were administered to other students and faculty at the school. In all, 32 of 172 students reported symptoms of diarrhea (100%), cramps (80%), nausea (51%), fever (29%), vomiting (26%), and bloody stools (14%). The food questionnaire clearly implicated milk as the common source, and a dose/response was evident (those drinking more milk were more likely to be ill). Investigation of the dairy supplying the milk showed that they vat pasteurized the milk at 135°F for 25 minutes rather than the required 145°F for 30 minutes. The dairy processed surplus raw milk for the school, and this milk had a high somatic cell count. Cows from the herd supplying the dairy had *C. jejuni* in their feces. This outbreak points out the variation in symptoms which may occur with campylobacteriosis and the absolute need to adhere to pasteurization time/temperature standards.

Yersinia Enterocolitica

1. *Name of the Organism: Yersinia enterocolitica* (and *Yersinia pseudotuberculosis*) *Y. enterocolitica,* a small rod-shaped, Gram-negative bacterium, is often isolated from clinical specimens such as wounds, feces, sputum and mesenteric lymph nodes. However, it is not part of the normal human flora. *Y. pseudotuberculosis* has been isolated from the diseased appendix of humans. Both organisms have often been isolated from such animals as pigs, birds, beavers, cats, and dogs. Only *Y. enterocolitica* has been detected in environmental and food sources, such as ponds, lakes, meats, ice cream, and milk. Most isolates have been found not to be pathogenic.
2. *Name of Disease:* Yersiniosis There are 3 pathogenic species in the genus Yersinia, but only *Y. enterocolitica* and *Y. pseudotuberculosis* cause gastroenteritis. To date, no foodborne outbreaks caused by *Y. pseudotuberculosis* have been reported in the United States, but human infections transmitted via contaminated water and foods have been reported in Japan. *Y. pestis,* the causative agent of "the plague," is genetically very similar to *Y. pseudotuberculosis* but infects humans by routes other than food.
3. *Nature of Disease:* Yersiniosis is frequently characterized by such symptoms as gastroenteritis with diarrhea and/or vomiting; however, fever and abdominal pain are the hallmark symptoms. *Yersinia* infections mimic appendicitis and mesenteric lymphadenitis, but the bacteria may also cause infections of other sites such as wounds, joints and the urinary tract.

4. *Infective dose:* Unknown. Illness onset is usually between 24 and 48 hours after ingestion, which (with food or drink as vehicle) is the usual route of infection.
5. *Diagnosis of Human Illness:* Diagnosis of yersiniosis begins with isolation of the organism from the human host's feces, blood, or vomit, and sometimes at the time of appendectomy. Confirmation occurs with the isolation, as well as biochemical and serological identification, of *Y. enterocolitica* from both the human host and the ingested foodstuff. Diarrhea is reported to occur in about 80% of cases; abdominal pain and fever are the most reliable symptoms. Because of the difficulties in isolating *yersiniae* from feces, several countries rely on serology. Acute and convalescent patient sera are titered against the suspect serotype of *Yersinia spp*. Yersiniosis has been misdiagnosed as Crohn's disease (regional enteritis) as well as appendicitis.
6. *Associated Foods:* Strains of *Y. enterocolitica* can be found in meats (pork, beef, lamb, etc.), oysters, fish, and raw milk. The exact cause of the food contamination is unknown. However, the prevalence of this organism in the soil and water and in animals such as beavers, pigs, and squirrels, offers ample opportunities for it to enter our food supply. Poor sanitation and improper sterilization techniques by food handlers, including improper storage, cannot be overlooked as contributing to contamination.
7. *Frequency of the Disease:* Yersiniosis does not occur frequently. It is rare unless a breakdown occurs in food processing techniques. CDC estimates that about 17,000 cases occur annually in the USA. Yersiniosis is a far more common disease in Northern Europe, Scandinavia, and Japan.
8. *Complications:* The major "complication" is the performance of unnecessary appendectomies, since one of the main symptoms of infections is abdominal pain of the lower right quadrant. Both *Y. enterocolitica* and *Y. pseudotuberculosis* have been associated with reactive arthritis, which may occur even in the absence of obvious symptoms. The frequency of such postenteritis arthritic conditions is about 2-3%. Another complication is bacteremia (entrance of organisms into the blood stream), in which case the possibility of a disseminating disease may occur. This is rare, however, and fatalities are also extremely rare.
9. *Target Populations:* The most susceptible populations for the main disease and possible complications are the very young, the debilitated, the very old and persons undergoing

immunosuppressive therapy. Those most susceptible to postenteritis arthritis are individuals with the antigen HLA-B27 (or related antigens such as B7).

10. *Food Analysis:* The isolation method is relatively easy to perform, but in some instances, cold enrichment may be required. *Y. enterocolitica* can be presumptively identified in 36-48 hours. However, confirmation may take 14-21 days or more. Determination of pathogenicity is more complex. The genes encoding for invasion of mammalian cells are located on the chromosome while a 40-50 MDal plasmid encodes most of the other virulence associated phenotypes. The 40-50 MDal plasmid is present in almost all the pathogenic *Yersinia* species, and the plasmids appear to be homologous.
11. *Selected Outbreaks:* 1976. A chocolate milk outbreak in Oneida County, N.Y. involving school children (first reported yersiniosis incident in the United States in which a food vehicle was identified). A research laboratory was set up by FDA to investigate and study *Y. enterocolitica* and *Y. pseudotuberculosis* in the human food supply. Dec. 1981 - Feb. 1982. *Y. enterocolitica* enteritis in King County, Washington caused by ingestion of tofu, a soybean curd. FDA investigators and researchers determined the source of the infection to be an non-chlorinated water supply. Manufacturing was halted until uncontaminated product was produced. June 11 to July 21, 1982. *Y. enterocolitica* outbreak in Arkansas, Tennessee, and Mississippi associated with the consumption of pasteurized milk. FDA personnel participated in the investigation, and presumptively identified the infection source to be externally contaminated milk containers.

Plesiomonas Shigelloides

1. Name of the Organism: *Plesiomonas shigelloides* This is a Gram-negative, rod-shaped bacterium which has been isolated from freshwater, freshwater fish, and shellfish and from many types of animals including cattle, goats, swine, cats, dogs, monkeys, vultures, snakes, and toads. Most human *P. shigelloides* infections are suspected to be waterborne. The organism may be present in unsanitary water which has been used as drinking water, recreational water, or water used to rinse foods that are consumed without cooking or heating. The ingested *P. shigelloides* organism does not always cause illness in the host animal but may reside temporarily as a transient, noninfectious member of the intestinal

flora. It has been isolated from the stools of patients with diarrhea, but is also sometimes isolated from healthy individuals (0.2-3.2% of population).It cannot yet be considered a definite cause of human disease, although its association with human diarrhea and the virulence factors it demonstrates make it a prime candidate.

2. *Nature of Acute Disease:* Gastroenteritis is the disease with which *P. shigelloides* has been implicated.
3. *Nature of Disease: P. shigelloides* gastroenteritis is usually a mild self-limiting disease with fever, chills, abdominal pain, nausea, diarrhea, or vomiting; symptoms may begin 20-24 hours after consumption of contaminated food or water; diarrhea is watery, non-mucoid, and non-bloody; in severe cases, diarrhea may be greenish-yellow, foamy, and blood tinged; duration of illness in healthy people may be 1-7 days. The infectious dose is presumed to be quite high, at least greater than one million organisms.
4. *Diagnosis of Human Illness:* The pathogenesis of *P. shigelloides* infection is not known. The organism is suspected of being toxigenic and invasive. Its significance as an enteric (intestinal) pathogen is presumed because of its predominant isolation from stools of patients with diarrhea. It is identified by common bacteriological analysis, serotyping, and antibiotic sensitivity testing.
5. *Associated Foods:* Most *P. shigelloides* infections occur in the summer months and correlate with environmental contamination of freshwater (rivers, streams, ponds, etc.). The usual route of transmission of the organism in sporadic or epidemic cases is by ingestion of contaminated water or raw shellfish.
6. *Relative Frequency of Disease:* Most *P. shigelloides* strains associated with human gastrointestinal disease have been from stools of diarrheic patients living in tropical and subtropical areas. Such infections are rarely reported in the U.S. or Europe because of the self-limiting nature of the disease.
7. *Course of Disease and Complications: P. shigelloides* infection may cause diarrhea of 1-2 days duration in healthy adults. However, there may be high fever and chills and protracted dysenteric symptoms in infants and children under 15 years of age. Extra-intestinal complications (septicemia and death) may occur in people who are immunocompromised or seriously ill with cancer, blood disorders, or hepatobiliary disease.
8. *Target Populations:* All people may be susceptible to infection. Infants, children and chronically ill people are more likely to experience protracted illness and complications.

9. *Food Analysis: P. shigelloides* may be recovered from food and water by methods similar to those used for stool analysis. The keys to recovery in all cases are selective agars which enhance the survival and growth of these bacteria over the growth of the background microflora. Identification following recovery may be completed in 12-24 hours.
10. *Selected Outbreaks: Literature references can be found at the links below.* Gastrointestinal illness in healthy people caused by *P. shigelloides* infection may be so mild that they do not seek medical treatment. Its rate of occurrence in the U.S. is unknown. It may be included in the group of diarrheal diseases "of unknown etiology" which are treated with and respond to broad spectrum antibiotics.

Streptococcus Spp

1. *Name of the Organism: Streptococcus* spp. The genus Streptococcus is comprised of Gram-positive, microaerophilic cocci (round), which are not motile and occur in chains or pairs. The genus is defined by a combination of antigenic, hemolytic, and physiological characteristics into Groups A, B, C, D, F, and G. Groups A and D can be transmitted to humans via food. Group A: one species with 40 antigenic types (*S. pyogenes*).Group D: five species (*S. faecalis, S. faecium, S. durans, S. avium,* and *S. bovis*).
2. *Nature of Acute Disease:* Group A: Cause septic sore throat and scarlet fever as well as other pyogenic and septicemic infections. Group D: May produce a clinical syndrome similar to staphylococcal intoxication.
3. *Nature of Disease:* Group A: Sore and red throat, pain on swallowing, tonsilitis, high fever, headache, nausea, vomiting, malaise, rhinorrhea; occasionally a rash occurs, onset 1-3 days; the infectious dose is probably quite low (less than 1,000 organisms). Group D: Diarrhea, abdominal cramps, nausea, vomiting, fever, chills, dizziness in 2-36 hours. Following ingestion of suspect food, the infectious dose is probably high (greater than 107 organisms).
4. *Diagnosis of Human Illness:* Group A: Culturing of nasal and throat swabs, pus, sputum, blood, suspect food, environmental samples. Group D: Culturing of stool samples, blood, and suspect food.
5. *Associated Foods:* Group A: Food sources include milk, ice cream, eggs, steamed lobster, ground ham, potato salad, egg salad, custard, rice pudding, and shrimp salad. In almost all cases, the foodstuffs were allowed to stand at room temperature for several hours between preparation and consumption. Entrance into the food is

the result of poor hygiene, ill food handlers, or the use of unpasteurized milk. Group D: Food sources include sausage, evaporated milk, cheese, meat croquettes, meat pie, pudding, raw milk, and pasteurized milk. Entrance into the food chain is due to underprocessing and/or poor and unsanitary food preparation.

6. *Relative Frequency of Disease:* Group A infections are low and may occur in any season, whereas Group D infections are variable.
7. *Course of Disease and Complications:* Group A: Streptococcal sore throat is very common, especially in children. Usually it is successfully treated with antibiotics. Complications are rare and the fatality rate is low. Group D: Diarrheal illness is poorly characterized, but is acute and self-limiting.
8. *Target Populations:* All individuals are susceptible. No age or race susceptibilities have been found.
9. *Food Analysis:* Suspect food is examined microbiologi-cally by selective enumeration techniques which can take up to 7 days. Group specificities are determined by Lancefield group-specific antisera.
10. *Selected Outbreaks: Literature references can be found at the links below.* Group A: Outbreaks of septic sore throat and scarlet fever were numerous before the advent of milk pasteurization. Salad bars have been suggested as possible sources of infection. Most current outbreaks have involved complex foods (i.e., salads) which were infected by a food handler with septic sore throat. One ill food handler may subsequently infect hundreds of individuals.

    Group D: Outbreaks are not common and are usually the result of preparing, storing, or handling food in an unsanitary manner.

7

Food Preservation through Pickling

Pickling is the process by which fresh fruits and vegetables are preserved and with the addition of salt, chilly and spices, a tasty preparation known as "Pickles" is made. Pickles are also good appetizers and digestive agents. There are several varieties of pickles and they are consumed throughout the year by people from all walks of life.

Unimaginable quantities of pickles are consumed round the year. On an average, each family consumes about 2 kgs of pickles every year. Pickled products truly add spice to meals and snacks.

The skilful blending of spices, sugar and vinegar with fruits and vegetables creates a crisp, firm texture and a pungent, sweet-sour flavour.

Although food markets today offer a wide variety of pickles and relishes, many homemakers like to make their own pickled products when garden vegetables and fresh fruits are in abundant supply.

Types of Pickled Products

Various types of pickled products can be made depending on the ingredients used and the methods of preparation. There are four general classes:

Brined Pickles or Fermented Pickles

These go through a curing process in a brine (salt and water) solution for one or more weeks. Curing changes the colour, flavour and texture of the product. If the product is a fermented one, the lactic acid produced during fermentation helps preserve the product. In brined products that are cured but not fermented, acid in the form of vinegar is added later to preserve the food.

Fresh Pack or Quick Process Pickles

These are covered with boiling hot vinegar, spices and seasonings. Sometimes, the product may be brined for several hours and then drained before being covered with the pickling liquid. These pickles are easy to prepare and have a tart flavour. Fresh pack or quick pickles have a better flavour if allowed to stand for several weeks after they are sealed in jars.

Fruit Pickles

These are prepared from whole or sliced fruits and simmered in a spicy, sweet-sour syrup made with vinegar or lemon juice.

Relishes

These are made from chopped fruits and vegetables cooked to desired consistency in a spicy vinegar solution. The level of acidity in a pickled product is as important to its safety as it is to its taste and texture. Never alter the proportions of vinegar, food or water in a recipe. Use only tested recipes. By doing so, you can help prevent the growth of *Clostridium botulinum,* the bacteria that produce a highly toxic poison in low acid foods.

Ingredients

Produce

Select tender vegetables and firm fruit. Pears and peaches may be slightly underripe for pickling. Always use a pickling variety of cucumbers. Do not expect good quality pickles if you use "table" or "slicing" cucumbers. Seed catalogues are a good source of information about cucumber varieties suitable for pickling. If you buy cucumbers, select unwaxed ones for pickling whole, because the brine solutions can't penetrate the wax. Use 1½-inch cucumbers for gherkins; 4-inch for dills. Oddshaped and more mature cucumbers should be used for relishes and bread-and-butter style pickles. For highest quality, plan to pickle the fruits or vegetables within 24 hours after they have been harvested. If the produce cannot be used immediately, refrigerate it, or spread it where it will be well-ventilated and cool.

This is particularly important for cucumbers because they deteriorate rapidly, especially at room temperature. Just before pickling, sort the fruits and vegetables and select the size best suited for the specific recipe. Wash well, especially around the stems. Soil trapped here can be a source of bacteria responsible for the softening of pickles.

Be sure to remove a 1/16-inch slice from the blossom end of the vegetables. The blossoms contain enzymes that also can cause softening. Do not use fruits and vegetables that show even slight evidence of mold. Proper processing kills potential spoilage organisms but does not destroy the off-flavour that may have already been produced by mold growth on the fruit or vegetables.

Salt

Pure granulated salt, such as "pickling" or "canning" salt should be used. It can be purchased from grocery, hardware or farm supply stores.

Other salts contain anti-caking materials that may make the brine cloudy. Do not alter salt concentrations in fermented pickles or sauerkraut. Proper fermentation depends on correct proportions of salt and other ingredients.

Vinegar

Use cider or white vinegar of 5-per cent acidity (50 grain). This is the range of acidity for most commercially bottled vinegars. Cider vinegar has a good flavour and aroma, but may darken white or light-coloured fruits and vegetables. White distilled vinegar is often used for onions, cauliflower and pears where clearness of colour is desired. Do not use homemade vinegar or vinegar of unknown acidity in pickling. Do not dilute the vinegar unless the recipe specifies. If a less sour product is preferred, add sugar rather than dilute the vinegar.

Sugar

Use white sugar unless the recipe calls for brown. White sugar gives a product a lighter colour, but brown sugar may be preferred for flavour. If you plan to use a sugar substitute, follow recipes developed for these products. Sugar substitutes are not usually recommended in pickling, as heat and/or storage may alter their flavour. Also, sugar helps to plump the pickles and keep them firm.

Spices

Use fresh whole spices for the best quality and flavour in pickles. Powdered spices may cause the product to darken and become cloudy. Pickles will darken less if you tie whole spices loosely in a clean white cloth or cheesecloth bag and then remove the bag from the product before packing the jars. Spices deteriorate and quickly lose their pungency in heat and humidity. Therefore, store any unused spices in an airtight container in a cool place.

Water

When brining pickles, hard water may interfere with the formation of acid and prevent pickles from curing properly. To soften hard water, simply boil it 15 minutes and let set for 24 hours, covered. Remove any scum that appears. Slowly pour water from the containers so the sediment will not be disturbed. Discard the sediment. The water is now ready for use. Distilled water can also be used in pickle making, but is more expensive.

Firming Agents

If good-quality ingredients are used and up-to-date methods are followed, lime and alum are not needed for crisp pickles. Soaking cucumbers in ice water for four to five hours prior to pickling is a safer method for making crisp pickles. If you choose to use firming agents, alum may be used to firm fermented cucumbers, but does not work with quick process pickles.

The calcium in lime does improve pickle firmness. If you choose to use lime, purchase food-grade pickling lime from your grocer's shelves. Do not use agricultural or burnt lime. Food-grade lime may be used as a lime-water solution for soaking fresh cucumbers 12 to 24 hours before pickling them. However, excess lime absorbed by the cucumbers must be removed to make safe pickles.

To remove excess lime, drain the lime-water solution, rinse and then re-soak the cucumbers in fresh water for one hour. Repeat the rinsing and soaking steps twice more. Failure to remove lime adequately may increase the risk of botulism.

Equipment

The right equipment prevents pickle failure and saves time and energy. Read each recipe completely to make sure you have the right equipment before you start to make pickled products.

There are a variety of utensils you need to make pickles and relishes. The one main point to remember is that pickles are acid or prepared with a heavy salt solution, check to make sure that you have suitable containers.

For fermenting or brining...use a crock or stone jar, unchipped enamel-lined pan, heavy foodgrade plastic container, or large glass jar or bowl. Jars and crocks need loose-fitting covers.

Use a heavy plate or large glass lid topped with a weight to keep vegetables below the surface of the brine. A foodgrade plastic bag containing 4 1/2 tablespoons of salt and 3 quarts of water or a glass jar filled with water makes a good weight.

Bricks or stone weights can impart an undesirable flavour. To heat pickling liquids...use unchipped enamel ware utensils, stainless steel, or aluminum. Never use anything that is galvanized, or copper, brass or iron. These metals react with the acids or salts to cause undesirable colour changes, and may form harmful compounds.

The pot should be large enough to simmer the ingredients without boiling over and, if the recipe calls for a rolling boil, it should have three times the capacity of the recipe. Use open kettles unless otherwise specified.

Use a boiling water bath canner to process jars after filling. Standard canning jars in good condition and new two-piece canning lids are preferred. However, mayonnaise or salad dressing jars and new two-piece canning lids may be used. Miscellaneous kitchen equipment includes: a teakettle, saucepans, spoons to stir and slotted spoons to lift, a ladle, funnel, standard measures, household scale, timer, tongs for lifting jars in and out of boiling water, bath canner and pot holders.

Containers and Weights for Fermentation

Pickles and sauerkraut can be fermented in large stoneware crocks, large glass jars or food-grade plastic containers. To determine if a plastic container is food-grade, check the label or contact its manufacturer. Or, line the questionable container with several thicknesses of food-grade plastic bags. Do not use aluminum, copper, brass, galvanized or iron containers for fermenting pickles or sauerkraut.

The container needs to be large enough to allow several inches of space between the top of the food and the top of the container. Usually a 1-gallon container is needed for each 5 pounds of fresh vegetables. After the vegetables are placed in the container and covered with brine, they must be completely submerged in the brine.

A heavy plate or glass lid that fits down inside the container can be used. If extra weight is needed, a glass jar(s) filled with water and sealed can be set on top of the plate or lid. The vegetables should be covered by 1 to 2 inches of brine. Another option for submerging the vegetables in brine is to place one food-grade plastic bag inside another and fill the inside bag with some of the pickling brine, in case the bags are accidentally punctured. Freezer bags sold for packaging turkeys are the right size for 5-gallon containers. Close the end securely. Then use this filled bag as the weight on top of the vegetables.

Equipment for Fresh Pack Pickles

Pickling liquids should be heated in a stainless steel, aluminum, glass or unchipped enamelware saucepan. Do not use copper, brass, galvanized or iron utensils.

These metals can react with acids or salts and cause undesirable colour changes in the pickles. For short-term brining or soaking, use crocks, saucepans or bowls made from stoneware, glass, stainless steel, aluminum or unchipped enamelware.

Except for the aluminum, the same containers can be used for soaking vegetables in lime. Lime pits aluminum containers and can cause an increased level of aluminum in the pickles.

Household Scales

Household scales will be needed if the recipes specify ingredients by weight. They are necessary in making sauerkraut to ensure correct proportions of salt and shredded cabbage. Processing equipment is needed for processing in a water bath canner.

Processing

Processing is necessary for all pickles and relishes to destroy the yeasts, moulds and bacteria that may cause the product to spoil and also to inactivate enzymes that could affect the colour, flavour and texture of the pickled product. As in all canning, a seal is necessary on the jar to prevent other microorganisms from entering.

Pickles and relishes are high-acid products. This acid may come from the large amount of vinegar added or, in the brined or fermented pickles, the acid is produced naturally during the fermentation process by lactic acid bacteria. Because they are high acid foods, they are processed in a boiling water bath canner. All canning jars should be washed in soapy water, rinsed well and then kept hot. Jars that will be processed for less than 10 minutes in a boiling water bath canner do need to be sterilized by boiling them for 10 minutes before filling. Jars processed in a boiling water bath canner for 10 minutes or more will be sterilized during processing.

Use new two-piece lids and follow the manufacturer's instructions for treating them. Carefully place the filled jars onto a rack in the canner containing hot water. The water should be deep enough to cover the jars by at least 1 inch. Cover the canner and bring water to a boil. Start counting processing time as soon as the water begins to boil. Process for the length of time specified in the recipe. Keep the water boiling. If no time is given, process the pickled product for at least 10 minutes.

On Guard Against Spoilage

Always be on the alert for signs of spoilage. Before opening a jar examine it closely. A bulging lid or leakage may mean the contents are spoiled. When a jar is opened, look for other signs of spoilage such as spurting liquid, disagreeable odor, change in colour or unusual softness, and mushiness or slipperiness of product. If there is even the slightest indication of spoilage, do not taste the contents. Dispose of the food so that it cannot be eaten by humans or animals.

Quick Process Pickles

Quick process pickles differ from fermented pickles because the *pickling* process uses acetic acid from vinegar rather than lactic acid from fermentation. Quick process pickles are ideal for those who want to make pickles, start to finish, in a few days. However, the flavour of fresh pack or quick process pickles is better if they are left to stand in sealed jars for several weeks.

Selection

Select tender vegetables. Always use a cucumber variety recommended for *pickling*. Refer to seed catalogs for information.

Catalogues often distinguish varieties best for pickling by labeling them for "table use" or for "slicing". If you buy cucumbers, select only unwaxed ones. For highest quality, plan to pickle the vegetables within 24 hours after they are picked. If the produce cannot be used immediately, refrigerate it, or temporarily store it in a cool, well-ventilated place. This is particularly important for cucumbers because they deteriorate rapidly, especially at room temperature.

Preparation

Sort the vegetables and select the size best suited for your recipe. Do not use vegetables showing evidence of mould. Wash vegetables well, especially around the stems. Soil trapped around the stem breeds bacteria that softens pickles. Because the blossom end of the vegetables contains enzymes that can cause softening, it should also be removed. Proper processing kills organisms that cause spoilage but does not destroy the off-flavour produced by mold growth on the vegetables.

Ingredients

Salt

Use canning salt, pickling salt, or kosher salt when making quick process pickles. Although iodized and non-iodized table salt may be used safely, the non-caking materials added to table salts may make the brine cloudy and the iodine may darken the pickles. Do not use flake salt as it varies in density. *Reduced-sodium salt may be used in quick process pickle recipes, but not in fermented pickle recipes.*

Vinegar

Use white distilled or cider vinegar of 4 per cent to 6 per cent acidity (40 to 60 grain). Use white vinegar for a lighter colour (especially when pickling onions and cauliflower).

Sweeteners

White granulated and brown sugars are the sweeteners most often used in pickling. White sugar will not add any colour to the product; however, brown sugar may add a distinctive flavour and colour. Do not use a sugar substitute unless the recipe has been developed for that product.

Firming Ingredients

Lime and alum are not needed for crisp pickles. Alum does not improve the firmness of quick process pickles. An excess of alum can cause digestive upsets. Lime may lower the pH of a pickled product, thus allowing harmful organisms to grow. For safe limed pickles, soak cucumbers in a mixture of 1 cup pickling lime (food grade) and 1/2 cup salt to 1 gallon of water. Do not inhale the lime water solution while mixing. Soak cucumbers in lime water for 12 to 24 hours, then remove the excess lime. Remove cucumbers from the lime solution, rinse, and soak for 1 hour in fresh water two more times. Handle carefully, as slices will be brittle. Drain well.

Spices

Use fresh whole spices for best quality and flavour in pickles. Powdered spices may darken and cloud the product. Pickles will darken less if you tie whole spices loosely in a clean white cloth or cheesecloth bag, then remove the bag from the pickling solution before packing the jars.

Processing

Use only recipes with tested proportions of ingredients. Sources of tested recipes include the United States Department of Agriculture (USDA), land-grant universities, and companies that manufacture pickling ingredients and canning equipment.

Precise measurements and tested recipes guarantee the quality and the safety of the product. Do not alter vinegar, salt, or water proportions in a recipe or use vinegar with an unknown acidity.

Sweet Gherkin Pickles

- 7 pounds cucumbers (1 1/2 inch or less)
- 1/2 cup canning or pickling salt
- 8 cups sugar
- 6 cups vinegar (5 per cent)
- 3/4 teaspoon turmeric
- 2 teaspoons celery seeds
- 2 teaspoons whole mixed pickling spice
- 2 cinnamon sticks
- 1/2 teaspoon fennel seed (optional)
- 2 teaspoons vanilla (optional).

Wash cucumbers. Cut 1/16-inch slice off blossom end and discard, leaving 1/4-inch of stem attached.

Place cucumbers in large container and cover with boiling water. Six to 8 hours later, and on the second day, drain and cover with 6 quarts of fresh boiling water containing 1/4 cup salt.

On the third day, drain and prick cucumbers with a table fork. Combine and bring to boil 3 cups vinegar, 3 cups sugar, turmeric, and spices. Pour over cucumbers. Six to 8 hours later, drain and save the pickling syrup. Add another 2 cups each of sugar and vinegar and reheat to boil.

Pour over pickles. On the fourth day, drain and save syrup. Add another 2 cups sugar and 1 cup vinegar. Heat to boiling and pour over pickles. Six to 8 hours later, drain and save pickling syrup. Add 1 cup sugar and 2 teaspoons vanilla and heat to boiling. Fill sterile pint jars with pickles and cover with hot syrup, leaving 1/2 inch headspace.

Quick Fresh Pack Dill Pickles

- 8 pounds of 3- to 5-inch pickling cucumbers
- 2 gallons water
- 1 1/4 cups canning or pickling salt (divided)
- 1 1/2 quarts vinegar (5 per cent)
- 1/4 cup sugar
- 2 quarts water
- 2 tablespoons whole mixed pickling spice
- 3 tablespoons whole mustard seed (1 teaspoon per pint jar)
- 14 heads of fresh dill (1 1/2 heads per pint jar)
- 4 1/2 Tablespoons dill seed (1 1/2 teaspoon per pint jar).

Wash cucumbers. Cut 1/16 slice off blossom end and discard, leaving 1/4-inch stem attached. Dissolve 3/4 cup salt in 2 gallons water.

Pour over cucumbers and let stand 12 hours. Drain. Combine vinegar, 1/2 cup salt, sugar, and 2 quarts water. Add mixed pickling spices tied in a clean white cloth. Heat to boiling. Fill jars with cucumbers. Add 1 teaspoon mustard seed and 1 1/2 heads fresh dill per pint. Cover with boiling pickling solution, leaving 1/2-inch headspace.

Pickling ChilE Peppers

Chile peppers are usually preserved in salt and vinegar.

Adding sugar or honey produces a more mellow tasting pickle.

Adding spices gives additional flavour:

- *Salt*: Used to extract moisture from some vegetables, which would otherwise dilute the vinegar and cause the pickle to ferment, and the vegetables to toughen. Use cooking salt rather than table salt; the latter contains a higher iodine content which can discoulor the vegetables (it makes them darker).
- *Vinegar*: Use bottled vinegar's (malt, distilled, wine, cider, spiced

etc.,); draught vinegar's are not strong enough. White distilled or cider vinegar's of 5 per cent acidity (50 grain) are recommended. Use cold vinegar for crisp vegetables and boiling vinegar for softer ones. The vinegar should cover the vegetables by at least an inch (2.5 cm). For a home-made spiced vinegar, boil 1oz (25 g) of mixed pickling spice in 1 pint (600 ml) of vinegar for 5 minutes. Strain when cold.

- *Spices*: Use whole spices, powdered ones will make the vinegar cloudy. Mixed pickling spice consists of equal amounts of stick cinnamon, allspice berries, cloves, mace and peppercorns. Extras can include root ginger, celery seeds, etc.
- *Pre-cooking*: Some recipes require the chile peppers to be blanched before pickling, some don't.
- *Pans*: Use unchipped enamel, aluminum or stainless steel pans. Copper, brass and iron pans will react with the vinegar, giving an off taste.
- *Sealing*: Jars must be sterilized and well sealed. Metal lids will corrode on contact with the vinegar. Kilner jars are recommended.
- *Maturing*: A minimum of 2 to 4 weeks maturing time is recommended. Crisp pickles will tend to soften after about 3 months.

Garlic Pickled Chile Peppers

- 500 g (1 lb) Cayenne peppers
- 2 tablespoons salt
- 1/4 teaspoon allspice
- 1/4 teaspoon celery seeds
- 1/2 teaspoon mustard seeds
- 3 cloves of garlic cut in halves
- 900 ml (1½ pints) vinegar (garlic vinegar if available)
- 1 tablespoon granulated sugar
- 8 peppercorns.

Wash the peppers and place into hot sterilized jars. Mix the allspice with the celery and mustard seeds. Pack into the jars. Add the garlic. Place the vinegar, sugar and peppercorns into pan and bring to the boil. Pour over the peppers and seal the jars. Makes about 1.5 kg (3 lb)

Fresh Pickled Jalapenos

Using fresh Jalapenos peppers, blanch for 3 minutes in boiling water.

To prevent collapsing, puncture each pepper with a needle. Add the following ingredients to a pint jar packed with the blanched peppers before cooling occurs:

- 1/4 medium sized Garlic clove
- 1/4 teaspoon of onion flakes
- 1 small bay leaf
- 1/8 teaspoon ground Oregano
- 1/8 teaspoon Thyme
- 1/8 teaspoon Marjoram
- 1 tablespoon vegetable oil.

Cover with boiling brine solution prepared as follows: mix together;

- 3 tablespoons sugar
- 9 tablespoons salt
- 2 pints water
- 2 pints 5% vinegar.

Close containers and process for 10 minutes in boiling water, then cool. Note that the jalapenos must be hot when brine solution is added.

Easy Pickled Peppers

- 1 - 1½ lb. fresh hot peppers (any kind you like)
- 1 good sized handful of cayenne peppers (optional-adds colour to mix)
- 1 lb package of peeled baby carrots
- 1 or 2 heads garlic—peel and separate cloves
- 1/4 cup salt
- 1/4 cup black peppercorns
- 1/4 cup whole coriander seeds
- White vinegar to cover.

Put the carrots on to boil in the vinegar. Stab each pepper with a paring knife.

After the carrots have cooked for about 10 minutes, add everything else to the pot. Simmer 5 (crisp) to 15 (soft) minutes depending on your taste.

Pour mixture into old mayonnaise jars or what ever else large glass jars you have on hand that have a cover. Cover (not too tight) and let cool for an hour or so. Then refrigerate. You can eat these right away but if you way for a few days the vegetables (carrots and garlic) will get hotter.

Grandma's Pickled Banana Peppers

- Banana peppers
- 1 tsp. salt
- 2 tbs. good olive oil in each jar
- Brine
- 3 qts. water
- 1 qt. white vinegar bring to boil.

Place peppers, salt, oil in jars, pour hot brine to the top of jar, seal jars, water bath for about 5 min. (When I water bath I only boil about 2 min). Peppers stay very crisp. Also you can put a clove or two of garlic in each jar if you wish.

Pickled Jalapenos

- 6 pounds jalapeno
- 5 cups vinegar
- 1 cup water
- 4 teaspoons pickling salt
- 2 tablespoons sugar
- 2 cloves garlic.

Wash peppers. If small peppers are left whole, slash 2-5 slits in each. Quarter large peppers. Blanch in boiling water. Flatten small peppers. Fill half-pint or pint jars leaving 1/2 inch head space. Combine and heat other ingredients to boiling and simmer 10 minutes.

Remove garlic. Pour hot pickling solution over peppers leaving 1/2 inch head space. Adjust lids. Use conventional boiling water canner processing. Process for 10 minutes at below 1000 feet, 15 minutes at 1,001 to 6,000 feet and 20 minutes above 6,000 feet.

Pickled Chiles

- 225g/8 oz. whole fresh chiles
- 350 ml/12 fl. oz/1 1/2 cups white wine vinegar with 1 teaspoon salt
- sprig of bay
- sprig of rosemary
- 4 gloves of garlic, peeled
- up to 350 ml/12 fl. oz/1 1/2 cups extra white wine vinegar.

Inspect the chiles for damage, discarding any that are bruised, lacerated or otherwise blemished. Snip off all but the base of their stems. Bring the vinegar and the chiles to a boil in a pan. Add the remaining ingredients except the extra vinegar and simmer for 6-8 minutes. With a clean spoon, transfer them to a jar previously sterilized with freshly boiled water. Pour in the pickling liquid with its herbs, top up with the additional vinegar to cover and allow to cool before sealing. The chiles will be ready within a month.

This quantity will fill a medium-sized jar with whole chiles, preserved with herbs and garlic and up to 675 ml/1½ pints /3 cups of vinegar.

Increase the vinegar quantity and dilute it with a little water if you want to preserve a larger quantity of chiles, or sweet red and yellow peppers, which should first have their caps, seeds and pithy membranes removed.

Dill Pickles

How Does Fermentation Preserve?

Brined pickles or fermented pickles go through a curing process in a brine (salt and water) solution for one or more weeks. Curing changes the colour, flavour and texture of the product.

If the product is a fermented one, the lactic acid produced during fermentation helps preserve the product. In brined products that are cured but not fermented, acid in the form of vinegar is added later to preserve the food.

Fermented Dill Pickles

Use the following quantities for each gallon capacity of your container.

Ingredients

- 4 pounds of 4-inch pickling cucumber
- 2 tablespoons dill seed or 4 to 5 heads fresh or dry dill
- 1/2 cup pure granulated salt ("pickling" or "canning" salt)
- 1/4 cup vinegar (5 per cent acidity)
- 8 cups water
- 2 cloves garlic (optional)
- 2 dried red peppers (optional)
- 2 teaspoons whole mixed pickling spices (optional).

Procedure

Wash cucumbers. Cut 1/16 inch slice off blossom end and discard. Leave 1/4 inch of stem attached. Place half of dill and spices on bottom of a clean, suitable container.

Add cucumbers, remaining dill and spices. Dissolve salt in vinegar and water and pour over cucumbers. Add suitable cover and weight. Store where temperature is between 70 and 75 °F for about 3 to 4 weeks while fermenting. Temperatures of 55 to 65 °F are acceptable, but the fermentation will take 5 to 6 weeks. Avoid temperatures above 80 °F, or pickles will become too soft during fermentation.

Fermenting pickles cure slowly. Check the container several times a week and promptly remove surface scum or mold. Caution: If the pickles become soft, slimy or develop a disagreeable odour, discard them. Fully fermented pickles may be stored in the original container for about 4 to 6 months, provided they are refrigerated and surface scum and moulds are removed regularly.

Fully fermented pickles are better stored by canning. To process them, pour the brine into a pan, heat slowly to a boil and simmer 5 minutes. Filter brine through paper coffee filters to reduce cloudiness, if desired. Fill jar with pickles and hot brine, leaving ½-inch headspace. Adjust lids and process pints for 10 minutes and quarts for 15 minutes in a boiling water bath. For the best sauerkraut, use firm heads of fresh cabbage. Shred cabbage and start kraut between 24 and 48 hours after harvest.

Ingredients

- 25 pounds cabbage
- 3/4 cup canning or pickling salt

Procedure

Work with about 5 pounds of cabbage at a time. Discard outer leaves. Rinse heads under cold running water and drain. Cut heads in quarters and remove cores. Shred or slice to a thickness of a quarter. Put cabbage in a suitable fermentation container. Add 3 tablespoons of salt.

Mix thoroughly, using clean hands. Pack firmly until salt draws juices from cabbage. Repeat shredding, salting and packing until all cabbage is in the container. Be sure it is deep enough so that its rim is at least 4 or 5 inches above the cabbage. If juice does not cover cabbage, add boiled and cooled brine (1½ tablespoons of salt per quart of water). Add plate and weights; cover container with a clean bath towel. Store at 70 to 75 °F while fermenting. At temperatures between 70 and 75 °F, kraut will be fully fermented in about 3 to 4 weeks; at 60 to 65 °F, fermentation may take 5 to 6 weeks.

At temperatures lower than 60°F, kraut may not ferment. Above 75 °F, kraut may become soft. If you weigh the cabbage down with a brine-filled bag, do not disturb the crock until normal fermentation is completed (when bubbling ceases). If you use jars as weight, check the kraut two to three times each week and remove scum if it forms.

Collard Kraut

Ingredients

- 5 gallons collards
- 1 cup canning or pickling salt.

Procedure

Rinse collards, chop or shred to desired consistency. Layer about 1 gallon of collards and sprinkle with 2 tablespoons salt in large glass jars, food-approved plastic containers or stoneware crocks. Then add more layers of collards and salt until container is full, leaving approximately 4 to 5 inches of space at top of container. Add water until it covers the chopped collards. The collards should be completely submerged in the brine. Add plate and weights; cover container with a clean bath towel. Store at 70°F for fermenting. At this temperature it will take approximately 3 to 4 weeks to ferment. If any scum forms above the plate or weight, remove it about 2 to 3 times a week. Taste in about two weeks. Allow collards to ferment until desired flavour is reached.

Storage

Fully fermented kraut may be kept tightly covered in the refrigerator for several months or it may be canned. *Hot Pack*: Bring kraut and liquid slowly to a boil in a large kettle, stirring frequently. Remove from heat and fill jars rather firmly with kraut and juices, leaving 1/2-inch headspace. Adjust lids and process pints for 10 minutes and quarts for 15 minutes in a boiling water bath.

Raw Pack: Fill jars firmly with kraut and cover with juices, leaving 1/2-inch headspace. Adjust lids and process 20 minutes for pints and 25 minutes for quarts in a boiling water bath.

Making Fermented Sauerkraut

Fermented or brined products, such as dill cucumbers or sauerkraut, cure for several weeks. Curing changes the colour, flavour, and texture of the product. Lactic acid produced during fermentation helps preserve the product. Satisfactory products can be obtained only when quality ingredients are used and proper procedures followed. Correct proportions of vegetables, salt, vinegar, and spices are essential.

Ingredients

- *Cabbage*: Cabbage composition differs with the variety and growing conditions. Mature, large-headed types, weighing 6 to 15 pounds with a solid, white interior, are the most desirable for kraut. Red cabbage may be used but makes an unattractive product.

- *Cucumbers*: Select fresh, firm unwaxed cucumbers about 4 inches long. Cucumbers deteriorate rapidly, especially at room temperature. For best results, use varieties developed for pickling.
- *Salt*: Use non-iodized salt; iodine can prevent the bacterial fermentation. Canning or pickling salt can be purchased at most supermarkets. Changing salt proportions or using a reduced-sodium salt substitute in fermented products is not recommended because the product will not ferment correctly.
- *Spices*: Use fresh, whole spices for best flavour in pickles.

Equipment

The fermentation equipment must be washed in hot sudsy water and rinsed well with very hot water before use.

Suitable Containers

A 1-gallon container is needed for each 5 pounds of fresh vegetables. A 5-gallon stone crock is the ideal size for fermenting about 25 pounds of fresh cucumbers or shredded cabbage.

Food-grade plastic and glass containers are excellent substitutes for stone crocks. Other 1- to 3-gallon, non-food grade plastic containers may be used if lined inside with a clean food-grade plastic bag. Be certain that foods contact only food-grade plastics. Do not use garbage bags, trash liners, or plastic buckets not meant for food use.

Covers and Weights

Cabbage and cucumbers must be kept 1 to 2 inches under brine while fermenting. Insert a dinner plate or glass pie plate inside the fermentation container. The plate must be slightly smaller than the container opening, yet large enough to cover most of the shredded cabbage or cucumbers. To keep the plate under the brine, weight it down with 2 to 3 sealed quart jars filled with water. Covering the container opening with a clean, heavy bath towel helps prevent contamination from insects and moulds. The plate also can be weighted down with a large food-grade plastic bag filled with 3 quarts of water containing 4½ tablespoons of salt. Be sure to seal the plastic bag. Freezer bags sold for packaging turkeys are suitable for use with the 5-gallon containers.

Scales and Utensils

Household scales will be needed if the recipe specifies ingredients by weight.

When heating pickling liquids, use unchipped enamelware, stainless steel, aluminum, or glass utensils. Other metals may cause undesirable colour changes in the pickles or form undesirable compounds.

Table. Fermenting Temperatures and Times

Temperature	Time	Comments
below 55°F	—	product may not ferment
55-65	5-6 weeks	
70-75	3-4	ideal temperature
above 80	—	product may sofetn or spoil

Sauerkraut in the Canning Jar

It is very difficult to make sauerkraut safely in this manner. Because of the high risk for spoilage, this procedure is not recommended.

Fermented Dill Pickles (by the Gallon)

Use the following quantities for each gallon capacity of your container:

- 4 pounds of 4-inch pickling cucumbers
- 2 tablespoons dill seed, or 4 to 5 heads fresh or dry dillweed
- 1/2 cup salt
- 1/4 cup vinegar (5 per cent)
- 8 cups water and one or more of the following ingredients:
  - 2 cloves garlic (optional)
  - 2 dried red peppers (optional)
  - 2 teaspoons whole mixed pickling spices (optional)

Wash cucumbers. Remove blossom end and discard, leaving 1/4 inch of stem attached.

Place half the dill and spices on bottom of a clean, suitable container. Add cucumbers, remaining dill, and spices.

Dissolve salt in vinegar and water and pour over cucumbers. Add suitable cover and weight. Ferment pickles. Check the container several times a week and promptly remove surface scum or mold. If the pickles become soft, slimy, or develop a disagreeable odour, discard them. Fully fermented pickles may be stored for about 4 to 6 months in the refrigerator. Canning fully fermented pickles is a better way to store them.

To can them, pour the brine into a pan, heat slowly to a boil, and simmer 5 minutes. Filter brine through paper coffee filters to reduce cloudiness, if desired. Fill jars with pickles and hot brine, leaving 1/2-inch headspace. Adjust lids and process as below, or use the *low-temperature pasteurization treatment.*

Table. Process Time for Fermented Dill Pickles in a Boiling-water Canner

		Process Time at Altitudes of	
Style of pack	Jar size	0-1,000 ft.	1,001-3,000 ft.
Raw	Paints	10 min.	15 min.
	Quarts	15	20

Low-temperature Pasteurization

The following treatment results in a better product texture but must be carefully managed to avoid possible spoilage. Place jars in a canner filled halfway with warm (120°F to 140°F) water. Add hot water to a level 1 inch above jars. Heat the water enough to maintain 180 to 185°F water temperature for 30 minutes. Check with a food thermometer to be certain that the water temperature is at least 180°F during the entire 30 minutes. Temperatures higher than 185°F may cause unnecessary softening of pickles. Use only for brined pickles when recipe indicates.

Preparing and Canning Fermented Foods and Pickled Vegetables

Selection of Fresh Cucumbers

- *Quantity*: An average of 14 pounds is needed per canner load of 7 quarts; an average of 9 pounds is needed per canner load of 9 pints. A bushel weighs 48 pounds and yields 16 to 24 quarts—an average of 2 pounds per quart.
- *Quality*: Select firm cucumbers of the appropriate size: about 1½ inches for gherkins and 4 inches for dills. Use odd-shaped and more mature cucumbers for relishes and bread-and-butter style pickles.

Low-temperature Pasteurization Treatment

The following treatment results in a better product texture but must be carefully managed to avoid possible spoilage. Place jars in a canner filled half way with warm (120 to 140°F) water. Then, add hot water to a level 1 inch above jars.

Heat the water enough to maintain 180 to 185°F water temperature for 30 minutes.

Check with a candy or jelly thermometer to be certain that the water temperature is at least 180°F during the entire 30 minutes. Temperatures higher than 185°F may cause unnecessary softening of pickles.

- *Caution*: Use only when recipe indicates.

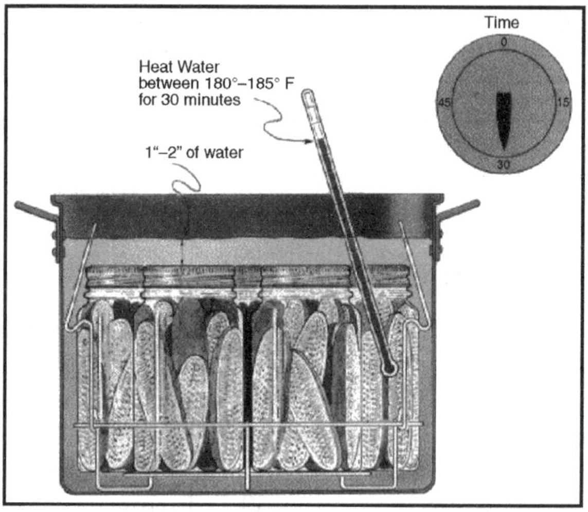

Suitable Containers, Covers, and Weights for Fermenting Food

A 1-gallon container is needed for each 5 pounds of fresh vegetables. Therefore, a 5-gallon stone crock is of ideal size for fermenting about 25 pounds of fresh cabbage or cucumbers. Food-grade plastic and glass containers are excellent substitutes for stone crocks. Other 1- to 3-gallon non-food-grade plastic containers may be used if lined inside with a clean food-grade plastic bag.

- *Caution*: Be certain that foods contact only food-grade plastics. Do not use garbage bags or trash liners. Fermenting sauerkraut in quart and half-gallon Mason jars is an acceptable practice, but may result in more spoilage losses.

Cabbage and cucumbers must be kept 1 to 2 inches under brine while fermenting. After adding prepared vegetables and brine, insert a suitably sized dinner plate or glass pie plate inside the fermentation container.

The plate must be slightly smaller than the container open-ing, yet large enough to cover most of the shredded cabbage or cucumbers. To keep the plate under the brine, weight it down with 2 to 3 sealed quart jars filled with water.

Covering the container opening with a clean, heavy bath towel helps to prevent contamination from insects and molds while the vegetables are fermenting. Fine quality fermented vegetables are also obtained when the plate is weighted down with a very large clean, plastic bag filled with 3 quarts of water containing 4-1/2 tablespoons of salt. Be sure to seal the plastic bag. Freezer bags sold for packaging turkeys are suitable for use with 5-gallon containers. The fermentation container, plate, and jars must be washed in hot sudsy water, and rinsed well with very hot water before use.

Salts Used in Pickling

Use of canning or pickling salt is recommended. Fermented and nonfermented pickles may be safely made using either iodized or noniodized table salt. However, noncaking materials added to table salts may make the brine cloudy. Flake salt varies in density and is not recommended for use. Reduced-sodium salts, for example "Lite Salt," may be used in quick pickle recipes, as indicated in this guide. The pickles may, however, have a slightly different taste than expected.

- *Caution*: Use of reduced-sodium salt in fermented pickle recipes is not recommended.

Fermented Foods

Pickled Horseradish Sauce

- 2 cups (3/4 lb) freshly grated horseradish
- 1 cup white vinegar (5%)
- 1/2 tsp canning or pickling salt
- 1/4 tsp powdered ascorbic acid.

Procedure

The pungency of fresh horseradish fades within 1 to 2 months, even when refrigerated. Therefore, make only small quantities at a time. Wash horseradish roots thoroughly and peel off brown outer skin. The peeled roots may be grated in a food processor or cut into small cubes and put through a food grinder. Combine ingredients and fill into sterile jars , leaving 1/4-inch headspace. Seal jars tightly and store in a refrigerator.

Marinated Whole Mushrooms

- 7 lbs small whole mushrooms
- 1/2 cup bottled lemon juice

- 2 cups olive or salad oil
- 2½ cups white vinegar (5%)
- 1 tbsp oregano leaves
- 1 tbsp dried basil leaves
- 1 tbsp canning or pickling salt
- 1/2 cup finely chopped onions
- 1/4 cup diced pimento
- 2 cloves garlic, cut in quarters
- 25 black peppercorns.

Procedure

Select very fresh unopened mushrooms with caps less than 1-1/4 inch in diameter. Wash. Cut stems, leaving 1/4 inch attached to cap. Add lemon juice and water to cover. Bring to boil. Simmer 5 minutes.

Drain mushrooms. Mix olive oil, vinegar, oregano, basil, and salt in a saucepan. Stir in onions and pimento and heat to boiling. Place 1/4 garlic clove and 2-3 peppercorns in a half-pint jar. Fill jars with mushrooms and hot, well-mixed oil/vinegar solution, leaving 1/2-inch headspace. Adjust lids and process.

Recommended Process Time for Marinated Whole Mushrooms in a Boiling-water Canner

		Process Time at Altitudes of			
Style of Pack	Jar Size	0–	1,001–	3,001–	Above
		1,000 ft	3,000 ft	6,000 ft	6,000 ft
Hot	Half-pints	20 min	25 min	30 min	35 min

Pickled Corn Relish

- 10 cups fresh whole kernel corn (16 to 20 medium-size ears), or six 10-ounce packages of frozen corn
- 2½ cups diced sweet red peppers
- 2½ cups diced sweet green peppers
- 2½ cups chopped celery
- 1½ cups diced onions
- 1½ cups sugar
- 5 cups vinegar (5%)
- 2½ tbsp canning or pickling salt

- 2½ tsp celery seed
- 2½ tbsp dry mustard
- 1½ tsp turmeric.

Procedure

Boil ears of corn 5 minutes. Dip in cold water. Cut whole kernels from cob or use six 10-ounce frozen packages of corn. Combine peppers, celery, onions, sugar, vinegar, salt, and celery seed in a saucepan.

Bring to boil and simmer 5 minutes, stirring occasionally. Mix mustard and turmeric in 1/2 cup of the simmered mixture. Add this mixture and corn to the hot mixture. Simmer another 5 minutes. If desired, thicken mixture with flour paste (1/4 cup flour blended in 1/4 cup water) and stir frequently. Fill jars with hot mixture, leaving 1/2-inch headspace. Adjust lids and process.

Recommended Process Time for Pickled Corn Relish in a Boilingwater canner

		Process Time at Altitudes of		
Style of Pack	Jar Size	0– 1,000 ft	1,001– 6,000 ft	Above 6,000 ft
Hot	Half-pints or Pints	15 min	20 min	25 min

Pickled Dilled Okra

- 7 lbs small okra pods
- 6 small hot peppers
- 4 tsp dill seed
- 8 to 9 garlic cloves
- 2/3 cup canning or pickling salt
- 6 cups water
- 6 cups vinegar (5%).

Procedure

Wash and trim okra. Fill jars firmly with whole okra, leaving 1/2-inch headspace. Place 1 garlic clove in each jar. Combine salt, hot peppers, dill seed, water, and vinegar in large saucepan and bring to a boil. Pour hot pickling solution over okra, leaving 1/2-inch headspace. Adjust lids and process.

Recommended Process Time for Pickled Dilled Okra in a Boilingwater Canner

		Process Time at Altitudes of		
Style of Pack	Jar Size	0–	1,001–	Above
		1,000 ft	6,000 ft	6,000 ft
Hot	Pints	10 min	15 min	20 min

Marinated Peppers

Bell, Hungarian, banana, or jalapeno

- 4 lbs firm peppers
- 1 cup bottled lemon juice
- 2 cups white vinegar (5%)
- 1 tbsp oregano leaves
- 1 cup olive or salad oil
- 1/2 cup chopped onions
- 2 cloves garlic, quartered (optional)
- 2 tbsp prepared horseradish (optional).

Note: It is possible to adjust the intensity of pickled jalapeno peppers by using all hot jalapeno peppers (hot style), or blending with sweet and mild peppers (medium or mild style).

For Hot Style: Use 4 lbs jalapeno peppers.

For Medium Style: Use 2 lbs jalapeno peppers and 2 lb sweet and mild peppers.

For Mild Style: Use 1 lb jalapeno peppers and 3 lbs sweet and mild peppers.

Procedure

Select your favourite pepper. If you select hot peppers, wear rubber or plastic gloves while handling them or wash hands thoroughly with soap and water before touching your face. Peppers may be left whole. Large peppers may be quartered. Wash, slash two to four slits in each pepper, and blanch in boiling water or blister in order to peel tough-skinned hot peppers.

Peppers may be blistered using one of the following methods: Oven or broiler method: Place peppers in a hot oven (400°F) or broiler for 6-8 minutes or until skins blister. Range-top method: Cover hot burner, either gas or electric, with heavy wire mesh.

Place peppers on burner for several minutes until skins blister. Allow peppers to cool. Place in pan and cover with a damp cloth. This will make peeling the peppers easier. After several minutes of cooling, peel each pepper. Flatten whole peppers.

Mix all remaining ingredients in a saucepan and heat to boiling. Place 1/4 garlic clove (optional) and 1/4 teaspoon salt in each half pint or 1/2 teaspoon per pint. Fill jars with peppers, add hot, well-mixed oil/pickling solution over peppers, leaving 1/2-inch headspace. Adjust lids and process.

Recommended Process Time for Marinated Peppers in a Boiling-water Canner

		Process Time at Altitudes of			
Style of Pack	Jar Size	0–	1,001–	3,001–	Above
		1,000 ft	3,000 ft	6,000 ft	6,000 ft
Raw	Half-pints or Pints	15 min	20 min	20 min	25 min

Pickled Bell Peppers

- 7 lbs firm bell peppers
- 3½ cups sugar
- 3 cups vinegar (5%)
- 3 cups water
- 9 cloves garlic
- 4½ tsp canning or pickling salt.

Procedure

Wash peppers, cut into quarters, remove cores and seeds, and cut away any blemishes. Slice peppers in strips. Boil sugar, vinegar, and water for 1 minute. Add peppers and bring to a boil. Place 1/2 clove of garlic and 1/4 teaspoon salt in each sterile half-pint jar; double the amounts for pint jars. Add pepper strips and cover with hot vinegar mixture, leaving 1/2-inch headspace. Adjust lids and process.

Recommended Process Time for Pickled Bell Peppers in a Boilingwater Canner

		Process Time at Altitudes of		
Style of Pack	Jar Size	0–	1,001–	Above
		1,000 ft	6,000 ft	6,000 ft
Hot	Half-pints or Points	5 min	10 min	15 min

Pickled Hot Peppers

- 4 lbs hot long red, green, or yellow peppers
- 3 lbs sweet red and green peppers, mixed
- 5 cups vinegar (5%)
- 1 cup water
- 4 tsp canning or pickling salt
- 2 tbsp sugar
- 2 cloves garlic.

Caution: Wear rubber gloves when handling hot peppers or wash hands thoroughly with soap and water before touching your face.

Procedure

Wash peppers. If small peppers are left whole, slash 2 to 4 slits in each.

Blanch in boiling water or blister in order to peel. Quarter large peppers. Cool and peel off skin. Flatten small peppers.

Fill jars, leaving 1/2-inch headspace. Combine and heat other ingredients to boiling and simmer 10 minutes. Remove garlic. Add hot pickling solution over peppers, leaving 1/2-inch headspace. Adjust lids and process.

Recommended Process Time for Pickled Hot Peppers in a Boiling-water Canner

	Process Time at Altitudes of			
Style of Pack	Jar Size	0–	1,001–	Above
		1,000 ft	6,000 ft	6,000 ft
Raw	Half-pints or Pints	10 min	15 min	20 min

Pickled Pepper-onion Relish

- 6 cups finely chopped onions
- 3 cups finely chopped sweet red peppers
- 3 cups finely chopped green peppers
- 1½ cups sugar
- 6 cups vinegar (5%), preferably white distilled
- 2 tbsp canning or pickling salt.

Recommended Process Time for Pickled Pepper-onion Relish in a Boiling-water Canner

		Process Time at Altitudes of		
Style of Pack	Jar Size	0–	1,001–	Above
		1,000 ft	6,000 ft	6,000 ft
Hot	Half-pints or Pints	5 min	10 min	15 min

Piccalilli

- 6 cups chopped green tomatoes
- 1½ cups chopped sweet red peppers
- 1½ cups chopped green peppers
- 2½ cups chopped onions
- 7½ cups chopped cabbage
- 1/2 cup canning or pickling salt
- 3 tbsp whole mixed pickling spice
- 4½ cups vinegar (5%)
- 3 cups brown sugar.

Procedure

Wash, chop, and combine vegetables with 1/2 cup salt. Cover with hot water and let stand 12 hours. Drain and press in a clean white cloth to remove all possible liquid. Tie spices loosely in a spice bag and add to combined vinegar and brown sugar and heat to a boil in a sauce pan. Add vegetables and boil gently 30 minutes or until the volume of the mixture is reduced by one-half. Remove spice bag. Fill hot sterile jars, with hot mixture, leaving 1/2-inch headspace. Adjust lids and process.

Bread-and-butter Pickles

- 6 lbs of 4- to 5-inch pickling cucumbers
- 8 cups thinly sliced onions (about 3 pounds)
- 1/2 cup canning or pickling salt
- 4 cups vinegar (5%)
- 4½ cups sugar
- 2 tbsp mustard seed
- 1½ tbsp celery seed
- 1 tbsp ground turmeric

- 1 cup pickling lime (optional) for use in variation below for making firmer pickles.

Procedure: Wash cucumbers. Cut 1/16-inch off blossom end and discard. Cut into 3/16-inch slices. Combine cucumbers and onions in a large bowl.

Add salt. Cover with 2 inches crushed or cubed ice. Refrigerate 3 to 4 hours, adding more ice as needed. Combine remaining ingredients in a large pot. Boil 10 minutes. Drain and add cucumbers and onions and slowly reheat to boiling. Fill pint jars with slices and cooking syrup, leaving 1/2-inch headspace.

Adjust lids and process as below or use low-temperature pasteurization treatment. Variation for firmer pickles: Wash cucumbers. Cut 1/1 6-inch off blossom end and discard. Cut into 3/16-inch slices. Mix 1 cup pickling lime and 1/2 cup salt to 1 gallon water in a 2- to 3-gallon crock or enamelware container.

Avoid inhaling lime dust while mixing the lime-water solution. Soak cucumber slices in lime water for 12 to 24 hours, stirring occasionally. Remove from lime solution, rinse, and resoak 1 hour in fresh cold water. Repeat the rinsing and soaking steps two more times. Handle carefully, as slices will be brittle. Drain well.

Recommended Process Time for Bread-and-butter Pickles in a Boiling-water Canner

		Process Time at Altitudes of		
Style of Pack	Jar Size	0–1,000 ft	1,001–6,000 ft	Above 6,000 ft
Hot	Pints or Quarts	10 min	15 min	20 min

Storage

After processing and cooling, jars should be stored 4 to 5 weeks to develop ideal flavour.

Variation

Squash bread-and-butter pickles. Substitute slender (1 to 1-1/2 inches in diameter) zucchini or yellow summer squash for cucumbers.

Quick Fresh-pack Dill Pickles

- 8 lbs of 3- to 5-inch pickling cucumbers
- 2 gals water
- 1½ cups canning or pickling salt (divided)

- 1½ qts vinegar (5%)
- 1/4 cup sugar
- 2 quarts water
- 2 tbsp whole mixed pickling spice
- about 3 tbsp whole mustard seed (1 tsp per pint jar)
- about 14 heads of fresh dill (1½ heads per pint jar)
- 4½ tbsp dill seed (1½ tsp per pint jar).

Procedure

Wash cucumbers. Cut 1/16-inch slice off blossom end and discard, but leave 1/4-inch of stem attached.

Dissolve 1/2 cup salt in 2 quarts water. Pour over cucumbers and let stand 12 hours. Drain.

Combine vinegar, 1/2 cup salt, sugar, and 2 quarts water. Add mixed pickling spices tied in a clean white cloth. Heat to boiling.

Fill jars with pickles. Add 1 tsp mustard seed and 1-1/2 heads fresh dill per pint. Cover with boiling pickling solution, leaving 1/2-inch headspace.

Adjust lids and process as below or use the low-temperature pasteurization treatment.

Recommended Process Time for Quick Fresh-pack Dill Pickles in a Boiling-water Canner

		Process Time at Altitudes of		
Style of Pack	Jar Size	0– 1,000 ft	1,001– 6,000 ft	Above 6,000 ft
Raw	Pints	10 min	15 min	20 min
	Quarts	15 min	20 min	25 min

Reduced-sodium Sliced Dill Pickles

- 4 lbs (3- to 5-inch) pickling cucumbers
- 6 cups vinegar (5%)
- 6 cups sugar
- 2 tbsp canning or pickling salt
- 1½ tsp celery seed
- 1½ tsp mustard seed
- 2 large onions, thinly sliced
- 8 heads fresh dill.

Procedure: Wash cucumbers. Cut 1/1 6-inch slice off blossom end and discard. Cut cucumbers in 1/4-inch slices. Combine vinegar, sugar, salt, celery, and mustard seeds in large saucepan. Bring mixture to boiling. Place 2 slices of onion and 1/2 dill head on bottom of each pint jar. Fill jars with cucumber slices, leaving 1/2-inch headspace. Add 1 slice of onion and 1/2 dill head on top. Pour hot pickling solution over cucumbers, leaving 1/4-inch headspace. Adjust lids and process.

Recommended Process Time for Reduced-sodium Sliced Dill Pickles in a Boiling-water Canner

		Process Time at Altitudes of		
Style of Pack	Jar Size	0–	1,001–	Above
		1,000 ft	6,000 ft	6,000 ft
Raw	Pints	15 min	20 min	25 min

Sweet Gherkin Pickles

- 7 lbs cucumbers (1½ inch or less)
- 1/2 cup canning or pickling salt
- 8 cups sugar
- 6 cups vinegar (5%)
- 3/4 tsp turmeric
- 2 tsp celery seeds
- 2 tsp whole mixed pickling spice
- 2 cinnamon sticks
- 1/2 tsp fennel (optional)
- 2 tsp vanilla (optional).

Procedure

Wash cucumbers. Cut 1/16-inch slice off blossom end and discard, but leave 1/4-inch of stem attached. Place cucumbers in large container and cover with boiling water.

Six to 8 hours later, and on the second day, drain and cover with 6 quarts of fresh boiling water containing 1/4-cup salt.

On the third day, drain and prick cucumbers with a table fork. Combine and bring to boil 3 cups vinegar, 3 cups sugar, turmeric, and spices.

Pour over cucumbers. Six to 8 hours later, drain and save the pickling syrup. Add another 2 cups each of sugar and vinegar and reheat to boil.

Pour over pickles. On the fourth day, drain and save syrup. Add another 2 cups sugar and 1 cup vinegar. Heat to boiling and pour over pickles. Drain and save pickling syrup 6 to 8 hours later.

Add 1 cup sugar and 2 tsp vanilla and heat to boiling. Fill sterile pint jars with pickles and cover with hot syrup, leaving 1/2-inch headspace.

Adjust lids and process as below, or use the low temperature pasteurization treatment.

Pickle Relish

- 3 qts chopped cucumbers
- 3 cups each of chopped sweet green and red peppers
- 1 cup chopped onions
- 3/4 cup canning or pickling salt
- 4 cups ice
- 8 cups water
- 2 cups sugar
- 4 tsp each of mustard seed, turmeric, whole allspice, and whole cloves
- 6 cups white vinegar (5%).

Procedure

Add cucumbers, peppers, onions, salt, and ice to water and let stand 4 hours. Drain and re-cover vegetables with fresh ice water for another hour. Drain again. Combine spices in a spice or cheesecloth bag. Add spices to sugar and vinegar. Heat to boiling and pour mixture over vegetables. Cover and refrigerate 24 hours. Heat mixture to boiling and fill hot into clean jars, leaving 1/2-inch headspace. Adjust lids and process.

14-day Sweet Pickles

Can be canned whole, in strips, or in slices.

- 4 lbs of 2- to 5-inch pickling cucumbers
  (If packed whole, use cucumbers of uniform size)
- 3/4 cup canning or pickling salt
  (Separated—1/4 cup on each of the 1st, 3rd, and 5th days)
- 2 tsp celery seed
- 2 tbsp mixed pickling spices
- 5½ cups sugar
- 4 cups vinegar (5%).

Procedure

Wash cucumbers. Cut 1/16-inch slice off blossom end and discard, but leave 1/4-inch of stem attached. Place whole cucumbers in suitable 1-gallon container. Add 1/4 cup canning or pickling salt to 2 quarts water and bring to a boil. Pour over cucumbers. Add suitable cover and weight. Place clean towel over container and keep the temperature at about 70°F. On the third and fifth days, drain salt water and discard. Rinse cucumbers and rescald cover and weight. Return cucumbers to container. Add 1/4 cup salt to 2 quarts fresh water and boil. Pour over cucumbers. Replace cover and weight, and re-cover with clean towel. On the seventh day, drain salt water and discard. Rinse cucumbers and rescald containers, cover, and weight. Slice or strip cucumbers, if desired, and return to container. Place celery seed

Safe Salsa Advice

Most salsa recipes combine onions, peppers and tomatoes with added acid (lemon juice, lime juice or vinegar). The type and amount of ingredients and preparation methods are important considerations in how salsa is canned. If salsa or other tomato-vegetable mixtures are not canned properly, they can support the growth of a potentially harmful bacterium, *Clostridium botulinum.*

To home-can salsa safely, it is extremely important that you follow a tested and approved recipe for home-canning. Food scientists tested recipes to make sure that the amount of acid combined with other ingredients is safe for home-canning, and that the processing times and methods will destroy harmful microorganisms.

The recipe ingredient amounts and processing methods must be followed to the letter.

If you enjoy creating your own salsa recipe or tweaking a recipe, plan to freeze it or store it for several weeks in the refrigerator and eat it fresh.

Home-Canned Salsa Guidelines

- Do not reduce the amount of lemon juice, vinegar or tomatoes.
- Do not add extra peppers, onion or garlic but you can reduce the amount of peppers, onion or garlic.
- You can substitute one type of pepper for another.
- Paste tomatoes contribute to a thicker salsa, while slicing tomatoes yield a thinner salsa. Do not drain or squeeze tomatoes to remove the liquid and juices or you will be removing some of the needed acids.
- Red, yellow or white onions may be substituted for each other.

- The amount of dry spices may be altered or deleted.
- For a stronger cilantro flavour, add fresh cilantro just before serving.
- Never add flour or cornstarch to salsa before canning because heat may not penetrate through the jar to destroy harmful microorganisms. Thicken salsas by adding tomato paste or using a paste tomato.

Freeze Peppers

If you have a lot of peppers, store them unwashed in a perforated plastic bag in the refrigerator up to one week. Before using, rinse with cool water. To freeze peppers, choose crisp, tender green, bright red or bright yellow pods. Wash, cut out stem, cut in half and remove seeds.

Peppers may be cut into ½-inch strips or rings. If the peppers will be used in cooked dishes, water-blanch halves for 3 minutes; strips or rings for 2 minutes. Cool promptly, drain and package, leaving ½-inch head space. Seal, label and freeze. If the peppers will be used in uncooked foods, package them raw, leaving no head space. Seal, label and freeze.

Sweet Corn

Sweet corn is an easy vegetable to freeze and produces a tasty, quality product when frozen properly. Harvested corn on the cob is at its optimum quality for 48 hours. When harvesting, do so early in the morning before the heat of the day. Once husked, the ears are blanched in boiling water for 4 minutes, followed by a complete cooling in ice water for about 8 minutes.

Drain and cut the kernels from the cob. An electric knife is a handy tool for cutting off the kernels. Package the corn in freezer containers, leaving ½-inch headspace. Seal and freeze at 0°F or below for best quality. Blanching, followed by chilling in ice water, are critical processes for producing quality frozen corn. The natural enzymes in corn need to be inactivated by blanching to prevent both loss of colour and nutrients, and flavour and texture changes.

Inadequately chilled corn may become mushy due to overcooking the starch. Can you freeze corn on the cob? Yes, it can be done, but with mixed results. Corn lovers are often disappointed with the sometimes mushy, rubbery texture and the cobby taste. It also takes up more space in the freezer. You can enjoy the great taste of summertime sweet corn all year long by following the simple, basic procedures for freezing vegetables.

Pickling: It's Not Just for Cucumbers

Vegetables from asparagus to zucchini can be home preserved by pickling. The key is to select a recipe from an approved source that is specifically designed for the vegetable you are pickling. Follow the directions carefully for a safe, high quality product.

Begin by selecting tender vegetables and plan to pickle within 24 hours of picking. Wash well and drain.

Green beans, carrots, and zucchini are raw packed; whereas asparagus is blanched and beets need to be pre-cooked in their skins for 30 minutes. Cider vinegar has a good flavour and aroma, but may darken white or light-coloured vegetables. White distilled vinegar is the preferred choice for pickling.

Follow the recipe recommendations for the ratio of vinegar to water; it does vary by the vegetable. Pickling or canning salt should be used as other salts contain anti-caking materials that may make the brine cloudy. Use fresh whole spices for the best quality and flavour.

Pickled vegetable recipes are developed for pint or ½-pint canning jars. To insure a safe home canned product, water bath process for 5 minutes to 30 minutes as stated in the recipe.

Sauerkraut

A grilled bratwurst topped with sauerkraut, a summertime treat! Sauerkraut can easily be made at home with its basic ingredients cabbage and salt. To make good kraut, begin by selecting disease-free, firm, mature heads of cabbage and begin cleaning and shredding the cabbage within 24 to 48 hours of harvest.

A kraut cutter is the traditional way to shred the cabbage, but a modern-day food processor works well. Canning or pickling salt draws out the cabbage juice so it can ferment. Be sure to use a tested recipe when making sauerkraut as the proportion of salt to cabbage is key to quality kraut.

Using too little salt softens the cabbage and yields a product lacking flavour. Too much salt delays the natural fermentation process.

Sauerkraut tips:

- For every 5 pounds of shredded cabbage, mix in 3 tablespoons of canning salt.
- Select a food-grade plastic pail or an old-fashioned crock in good condition to hold the cabbage. Do not use a metal container.
- Cover the cabbage with a heavy plate, or a salt water brine-filled food grade plastic bag to exclude air.

- Store the container at 68° to 74°F while fermenting. The sauerkraut should be ready in 3 to 4 weeks.
- Sauerkraut may be canned and processed in a boiling water bath or frozen.

Half-gallon Jars

Half-gallon jars are hard to find but they are out there. The only home food preservation foods USDA and manufacturers recommend for half-gallon jars are grape juice or apple juice. Grape juice and apple juice are the only high acid juices safe for these large jars.

Currently, there is no research for safe processing times for other juices or products in half-gallon jars. The density of the product, water content and acid levels are the primary variables that affect safe processing times. At this time, it is not possible to determine a safe formula for other foods in large jars using the boiling water bath canner.

Pickling with salts, spices and oil

There will rarely be any house in India where pickles are not eaten. Recall the pickles that you ate this season. Do you know how salt, spices and oil help in preserving the pickles?

We know that every foodstuff has some amount of water in it, which helps micro-organisms to grow. When salt and spices are added, they draw out the water from the foodstuff, thus micro-organisms cannot grow. Moreover, they improve the flavour of the food being preserved. Spices such as mustard, pepper and *hing* (asafoetida) are used in pickling.

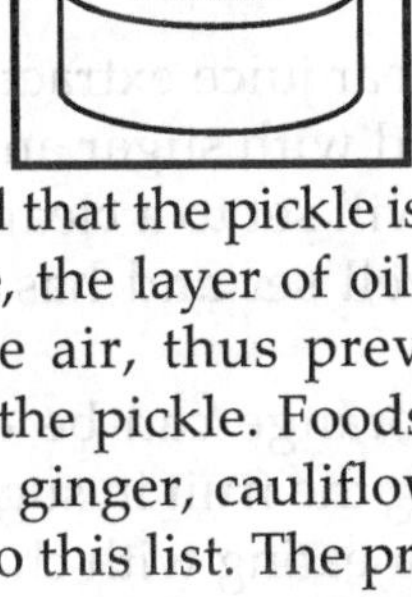

You must have observed that the pickle is usually covered with a layer of oil. Why? This is because, the layer of oil prevents the foodstuff from coming in contact with the air, thus preventing the entry of micro-organisms, which can spoil the pickle. Foodstuffs that can be pickled are lemon, mango, amla, carrot, ginger, cauliflower red/green chilli, etc. You can surely add many more to this list. The process of preparing a pickle is simple.

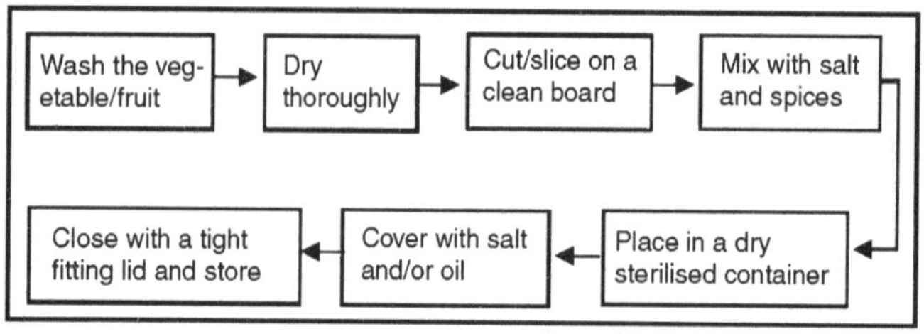

Here is a receipe for sweet lemon pickle. You can prepare other pickles in the same way as explained above.

Sweet Lemon Pickle

Ingredients	*Quantity*
Lemon	1 kg
Salt	150 g
Black salt	100 g
Cloves (powdered)	2 No.
Ajwain	40 g
Hing	1 g
Black pepper powder	40 g
Sugar	200 g

Method

Select round, fully mature and juicy lemons. Wash well and dry. Cut each lemon into 8 pieces. Put in a clean jar. Add all the spices and sugar. Shake well. Keep the pickle in sun for 10-12 days, shaking it occasionally.

Making jam, jellies and murabbas

You can make jam by boiling the fleshy part (pulp) of the fruit with sufficient quantity of sugar to a thick mixture.

You can make jams with apple, mango, papaya, pineapple, raw mango, etc. Jellies are prepared by boiling the fruit with or without water and straining it.

Then the strained and clear juice extract containing pectin (helps to set the jelly) and acid is mixed with sugar and the mixture is boiled. It is boiled to a stage at which it will set to a clear jelly-like substance. A well made jelly is transparent, well set and has the original flavour of the fruit.

Guava, pineapple and orange can be easily converted into jelly. Murabbas are made by boiling the fruit in sugar solution, till they become soft. Murabbas are popular for eating with chapatis, puris, etc. Let us now learn how to make a mixed fruit jam and amla murabba.

Select cucumbers of the appropriate size, about 1-1/2 inches for gherkins and 5 inches for dills. Use odd-shaped and more mature cucumbers for relishes and bread-and-butter style pickles. All vegetables should be fresh, firm and free of spoilage. Use canning or pickling salt. The noncaking material added to other salts could make the brine cloudy. Flake salt varies in density, so don't use it to make pickled and fermented foods. The salt used in making fermented sauerkraut and brined pickles not only provides characteristic flavour but also is vital to safety and texture. In fermented foods, salt favours the growth of desirable bacteria while inhibiting the growth of others.

Caution

Do not cut back on the salt when making sauerkraut or fermented pickles. White granulated and brown sugars are the sweeteners most often used in pickling. Corn syrup and honey, unless called for in reliable recipes, may produce undesirable flavours. Use white distilled and cider vinegars of 5 per cent acidity (50 grain). White vinegar is usually preferred when light colour is desirable. Never use homemade vinegar when making pickles.

8

Preservation: Canning, Freezing and Drying

Preserving Foods

Unless food is preserved in some manner, it begins to spoil soon after it is harvested or slaughtered. The main methods of food preservation that will keep food safe are canning, freezing and drying.

Preserving food at home means having:

- An abundant supply of a variety of foods when the fresh products aren't readily available;
- Specialties such as strawberry-fig preserves or green tomato relish that can't always be purchased;
- The satisfaction of preserving foods yourself.

However, preserving food at home may not save you money, depending on costs of buying or raising the food, the needed equipment, energy and time.

Canning

Canning is the process in which foods are placed in jars or cans and heated to a temperature that destroys microorganisms and inactivates enzymes. This heating and later cooling forms a vacuum seal. The vacuum seal prevents other microorganisms from re-contaminating food within the jar or can. High-acid foods such as fruits and acidified tomatoes can be processed or "canned" in boiling water, while low-acid vegetables and meats must be processed in a pressure canner at 240°F.

Pickling

Pickling is another form of canning. Pickled products have an increased acidity that makes it difficult for most bacteria to grow. Pickled products are also heated in jars at boiling temperatures to destroy other microorganisms present and form a vacuum in the jar.

Jams and Jellies

Jams and jellies have a very high sugar content. The sugar binds with the liquid present making it difficult for microorganisms to grow. To prevent surface contamination after the product is made and possible yeast or mold growth jams and jellies are canned, frozen or refrigerated.

Freezing

Freezing reduces the temperature of the food so that microorganisms cannot grow, yet many still live.

Enzyme activity is slowed down but not stopped during freezing:

- *Enzymes in Vegetables:* These must be inactivated by blanching to prevent loss of colour, flavour and nutrients. The vegetable is exposed to boiling water or steam for a specified time and then quickly cooled in ice water to prevent cooking. Blanching is essential for top quality frozen vegetables and helps to destroy microorganisms on the surface of vegetables.
- *Enzymes in Fruits:* These can cause browning and loss of vitamin C, and are controlled by the addition of ascorbic acid.

Drying

Drying removes most of the moisture from foods. Thus microorganisms cannot grow and enzyme action is slowed down. Dried foods should be stored in airtight containers to prevent moisture from rehydrating the products and allowing microbial growth.

Unsafe Canning Methods

Open-kettle Method

This outdated method of canning is now considered unsafe. In this method, foods were heated in a kettle, then poured into jars and a lid was placed on the jar. No processing was done. With this method there was often spoilage because bacteria, yeast and molds that contaminated the foods when the jars were filled were not killed by further processing. The growth of these microorganisms, in addition to spoiling the food, often caused any lids that did seal to later come unsealed. This method resulted in a very real danger of botulism.

Steam Canning

This is a newer method of canning that is not considered safe at this time. The jars are heated by steam. However, safe processing times have not been developed and steam canners are NOT recommended for either high-or low-acid foods. Low-acid foods canned in these canners are potentially deadly because of possible botulism contamination. Also, both low-and high-acid foods are often underprocessed and therefore could spoil.

Other Unsafe Methods

Canning food in the microwave oven, electric ovens, slow cookers or crock pots can be extremely dangerous, especially with low-acid foods, and is not recommended. So-called canning powders are useless as preservatives and do not replace the need for proper heat processing.

Adjustments for Altitudes

As altitude increases, water boils at lower temperatures. Because the lower temperatures are less effective for killing bacteria, processing time must be increased for boiling water bath canning at higher altitudes. For pressure canning, the pressure is increased.

The directions for canning foods are usually for an altitude of 0 to 1000 feet. If you are canning at an altitude over 1000 feet, check for altitude adjustments needed for canning each type of food. Ask your county extension agent to help you determine your altitude or check with your local airport.

On Guard against Spoilage

Don't taste or use food that shows any kind of spoilage! Look closely at all jars before opening them. A bulging lid or leaking jar is a sign of spoilage. When you open the jar, look for other signs, such as spurting liquid and off-odours or mold.

Spoiled canned foods should be discarded in a place where they will not be eaten by humans or pets. Spoiled meats, seafood and low-acid vegetables, should be detoxified to destroy any poisons that might be present before they are discarded.

To detoxify canned low-acid foods that have been spoiled, carefully remove the lid from the jar. Be extremely careful when following these directions not to spread or come in contact with suspect food or liquid.

Place the open jar(s) of food and the lids in a saucepot. Add enough hot water to cover the jar(s). Boil for 30 minutes and then cool. Drain water and dispose of food and lids. The jars may be reused. Improperly canned low-acid foods can contain botulinum toxin without showing signs of spoilage.

Low-acid foods are considered improperly canned and not safe if any of the following are true:

- The food was NOT processed in a pressure canner.
- The gauge of the canner was inaccurate.
- Up-to-date researched processing times and pressures were NOT used for the size of the jar, style of pack and kind of food being processed.

- Ingredients were added that were NOT in an approved recipe.
- Proportions of ingredients were CHANGED from the original approved recipe.
- The processing time and pressure were NOT correct for the altitude at which the food was canned.

Because improperly canned low-acid foods can contain botulinum toxin without showing signs of spoilage, they should also be detoxified and then discarded. Surfaces that come in contact with spoiled or questionable food should be cleaned with a solution of 1 part chlorine bleach to 5 parts water. Wet the surface with this solution and let stand 5 minutes before rinsing.

Home Canning Equipment

Canning is the process in which foods are placed in jars or cans and heated to a temperature that destroys micro-organisms and inactivates enzymes. This heating and later cooling forms a vacuum seal. The vacuum seal prevents other microorganisms from recontaminating the food within the jar. High-acid foods such as fruits and tomatoes can be processed or "canned" in a boiling water canner while low-acid vegetables and meats must be processed in a pressure canner at 240 °F.

Water Bath Canner

A water bath canner is a large covered cooking pot with a rack. Any large metal container may be used as long as it is deep enough for 1 inch of briskly boiling water to cover the jars.

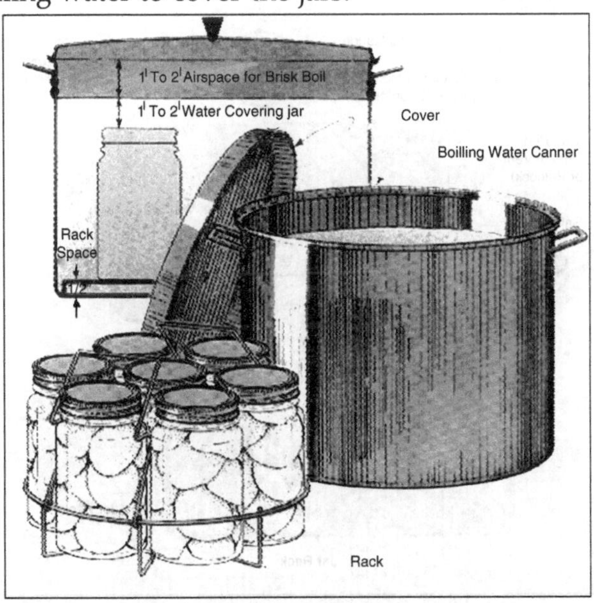

Fig. Boiling Water Bath Canner.

The diameter of the canner should be no more than 4 inches wider than the diameter of your stove's burner to ensure proper treatment of all jars. Using a wash kettle that fits over two burners is not recommended because the middle jars do not get enough heat. For an electric range, the canner must have a flat bottom.

The canner must have a tight-fitting lid and a rack. The rack keeps the jars from touching the bottom of the canner and allows the water to circulate freely under the jars. If the rack has dividers, jars will not touch each other or fall against the sides of the canner during processing. A deep pressure canner can be used as a boiling water bath canner. Just be sure there is enough space above the jars to allow for 1 inch of briskly boiling water. Place the lid loosely on the canner—don't fasten it. Leave the vent wide open, so that steam escapes and pressure does not build up inside.

Pressure Canner

A pressure canner is a specially-made heavy pot with a lid that can be closed to prevent steam from escaping. The lid is fitted with a vent, a dial-or weighted-pressure gauge, and a safety fuse. Newer models have an extra coverlock as an added precaution. It may or may not have a gasket. The pressure canner also has a rack. Because each type of a canner is different, be sure to read the directions for operating your canner and keep them for future reference.

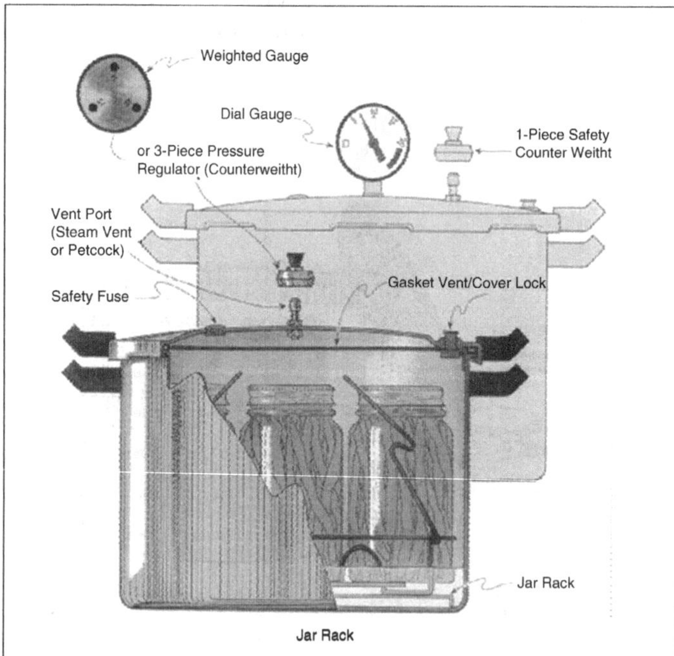

Fig. Pressure Canner.

Vent or Petcock

This is a short, hollow pipe that sticks up above the canner lid. When open, it allows air and steam to escape from the canner.

When closed, it holds the steam inside. On newer canners the vent is closed or opened using a separate pressure regulator weight. On older canners, the vent may be closed using a valve or screw that you can turn.

Pressure Gauge

This registers the pressure inside the canner. A dial gauge will actually show the temperature and/or pressure inside the canner.

The weighted gauge will rock gently or make a "jiggling" noise periodically to show that correct pressure is being maintained. Read the manufacturer's instructions to see how often the weight should rock or jiggle. Some canners have a three-piece weighted gauge that can regulate 5, 10 or 15 pounds of pressure. For 10 pounds of pressure, one piece of the weight is left off. Dial gauges and weighted gauges actually register slightly different pressures at their same settings. Because of this, dial gauges are operated at 11 pounds pressure up to an altitude of 2,000 feet. At altitudes over 2,000 feet, corrections must be made for dial-gauge canners. Weighted-gauge canners can be operated at 10 pounds pressure up to an altitude of 1,000 feet. At altitudes over 1,000 feet, correction must be made for weighted-gauge canners.

Gasket

This is a rubber or rubber-like compound that helps seal the edges of the canner and lid to prevent steam from escaping. Gaskets may be removable for cleaning or replacement.

Safety Checks for Pressure Canners

For safe operation, the vent, safety valve and edges of the lid and canner must be clean at all times. To clean the vent, draw a string or narrow strip of cloth through the opening.

The dial gauge on a canner should be checked for accuracy yearly. Check with your county Extension agent well in advance of each canning season for instructions on how this can be done. If the gauge is off more than 2 pounds at the pressure used for your altitude, replace it. Make pressure adjustments for the difference up to 2 pounds. Follow the manufacturer's directions for care of the sealing edges of your canner.

Pressure Saucepans and other Unsafe Models

Small pressure saucepans are not recommended for home canning. Also, outmoded and potentially unsafe pressure canners should not be used. Compare old canners with newer models to be sure that what you have is actually a pressure canner and not an old sterilizer or steamer. Before using an old canner, make sure all parts have been checked and are working properly. Buying an old second-hand canner may not be a bargain. Sometimes replacement parts are no longer being made.

Cleaning and Storing Canners

After use, clean your canner, being careful not to immerse the dial gauge on your pressure canner, if it has one. Dry the canner and store it with crumpled newspapers or paper towels in the bottom and around the rack. This will help absorb moisture and odours. Place the lid upside down on the canner. Never put the lid on the canner and seal it.

The darkened surface on the inside of an aluminum canner can be cleaned by filling it above the darkened line with a mixture of 1 tablespoon cream of tartar to each quart of water. Place the canner on the stove, heat water to a boil, and boil covered until the dark deposits disappear. Sometimes stubborn deposits may require the addition of more cream of tartar. Empty the canner and wash it with hot soapy water, rinse and dry.

Canning Jars and Lids

Mason-type jars specifically designed for home canning are best. Commercial mayonnaise jars may not seal and may break, especially in a pressure canner.

Old antique glass canning jars of any type are prized collector's items. These jars are often too brittle to withstand the heat treatment involved in canning and have a great likelihood of breaking during heat processing. They are best for uses other than canning.

Canning jars come in a variety of sizes from half-pint to half-gallon jars. Pint and quart jars are the most commonly used sizes. Processing times have not been developed for many foods in half-pint, 12-ounce or 1½ pint jars. If the recipe does not specify processing in one of these jars, process half-pint and 12-ounce jars for the same time as pints. Process 1½ pints for the same time as quarts. For jellied fruit products only, pint and 12-ounce jars can be processed for the same time as half-pint. Half-gallon canning jars are recommended only for very acid juices. Jars also come in both the regular and wide-mouth styles. If properly used, jars may be reused indefinitely.

Two-piece Lids

Most of the canning jars sold today have two-piece, self-sealing lids. This type consists of a flat metal disc that has a sealing compound around the outer edge and a separate metal screw band.

The lid is used only once; the screw band may be used over and over, unless it rusts. Do not use any old, dented or deformed lids or those with gaps or flaws in the sealing compound. These may not seal. Lids should be good for at least five years after manufacture. Never reuse lids from commercially canned foods for home canning.

Zinc Lids and Bail-type Jars

These used a rubber ring that fit on the jar's sealing ledge, but are no longer recommended and are not being manufac-tured. Some new imported bail-type jars are available in this country.

Many of these are not heat tempered and while these jars come with one rubber ring, no replacement rings are available for reuse. These jars are be tter used for food storage and decorative purposes than for home canning.

Additional Canning Utensils

The following items are helpful for home canning:

- A jar lifter, essential for easy removal of hot jars from canner.
- A jar filler or funnel helps in packing of small food items into canning jars.
- A bubble freer for removing air bubbles from the jars. This could be any plastic knife or spatula. Metal objects should not be used because they can scratch the glass. Making the jar more susceptible to breakage.
- A lid wand has a magnet on the end to help remove treated lids from hot water.
- Clean cloths for wiping jar rims and general cleanup.
- Knives for product preparations.
- Timer or clock to determine end of processing time.
- Clean towels.
- Cutting board.

A number of other home canning accessories such as corn cutters, apple slicers, decorative labels and special canning spoons are available. Some of these items may simplify the procedures but are not essential. Jar lid tighteners and wrenches are not recommended.

Home Freezing Equipment

The type of freezer to purchase will depend on family size, whether fresh produce or large quantities of meat will be frozen, available floor space, and efficiency and defrosting features preferred. The standard capacity of a freezer is about 35 pounds of frozen food per cubic foot of useable space. Families that freeze garden produce should allow six cubic feet of freezer space per person. If other methods of food preservation are used, allow three cubic feet per person. Additional space is needed if large amounts of meat are kept in the freezer.

Types of Freezers Available

There are three types of freezers on the market: upright, chest and refrigerator-freezer combinations. The upright and refrigerator-freezer are available as manual-defrost or frost-free models. Though less convenient, manual-defrost freezers are more cost-efficient. They also maintain higher quality food than frostless models because they don't have a fan running to remove the moisture that would turn to frost. The constant removal of moisture from the freezer could cause freezer burn in improperly wrapped food. Frost-free chest freezers are not available, but frost builds up in them less readily.

Upright Freezers

These appliances have the same general shape and appearance as home refrigerators. They have one or two outside doors and from three to seven shelves for storing food. Freezers of this type are popular due to their convenience, the small floor space they require and the ease with which food may be put in or removed. However, more cold air escapes each time the door is opened.

Chest Freezers

Freezers of this type require more floor area than the uprights but are more economical to buy and operate. These freezers lose less cold air each time they're opened. Make sure this type of freezer is equipped with sliding or lift-out baskets to permit easy loading and removal of food.

Refrigerator–Freezer Combination

This is a single appliance with one or two doors. It has one compartment for frozen foods and another for refrigerated foods. The freezing compartments may be above, below or to one side of the refrigerated area. If selecting this type, be certain that the freezer is a true freezer and not just a freezing compartment.

Care of the Freezer

Regardless of the type of freezer selected, it should be placed in a convenient, cool, dry and well-ventilated place; never place it by the stove, water heater or in the sun. This would make it more difficult to maintain a temperature of 0°F or lower. Do not push the freezer flush against a wall. Leave space for air circulation and cleaning. Be sure the freezer sits level.

Defrosting Freezers

Manual-defrost freezers need defrosting at least once a year or when there is more than one-fourth inch of frost over a large area of the freezer.

Accumulated freezer frost reduces storage space and increases operating costs. Defrosting should be scheduled when the food inventory is relatively low and defrosting can be completed within one to two hours. A manual-defrost model should be disconnected from the electrical supply before defrosting. Frozen packages should then be placed in large cardboard cartons or insulated ice chests. With a cardboard carton, several layers of newspapers may be used for extra insulation. Clean the freezer as quickly as possible, following your manufacturer's instructions. A few manufactures recommend placing pans of hot water in the freezer and closing it. Then, remove the frost as it loosens and replace the water as it cools.

Make sure the freezer is completely cool before restarting it. Other manufactures do not recommend using pans of hot water because in their freezers, refrigerator pressure could build up in the evaporator, making restarting the freezer difficult.

These manufactures recommend allowing the frost to thaw naturally or with the aid of a fan. Place towels in the bottom of the freezer to catch water and frost.

The loose frost can be removed using a wooden or plastic scrapper. When all the frost has been removed, sponge out the interior with a cleaning solution made of one tablespoon of baking soda per quart of water.

Sponge with clean water and dry with an absorbent cloth. Turn the freezer on and close the door to allow the freezer to become chilled before returning the food. If food packages are frosty, scrape or wipe them to remove frost or moisture before placing the food back in the freezer in an organized manner. Mark these packages for first use.

Care of Frost-free Freezers

A frost-free freezer does not need defrosting. However, it should be cleaned out once a year or more often if dirt or food residues are visible. Turn off the power source. Empty the freezer, wipe it with a baking soda solution, rinse, towel it dry and then replace the food.

Removing Odours

If food has spoiled in a freezer because of a power failure or some other reason, undesirable odours can develop. To eliminate the odour, remove the food and wash the inside of the freezer with one tablespoon of baking soda in a quart of tap water or with one cup of vinegar in a gallon of tap water. Let surface dry.

If the odour still persists, use activated charcoal. This type of charcoal is extra dry and absorbs odours more quickly than cooking-type charcoal. It can be purchased at a drug store or pet supply store. To use it, unplug the freezer. Put the charcoal in pans or on paper in the bottom of the freezer for several days.

If the odour remains, put in new charcoal. When the odour is gone, rinse and dry the inside of the freezer. Turn on the freezer and it is ready for food. When odour gets into the freezer's insulation, write the company for any suggestions it may have for solving the problem. However, sometimes, there is nothing than can be done.

Containers for Freezing

Foods for your freezer must have proper packaging materials to protect flavour, colour, moisture content and nutritive value from the dry climate of the freezer. Do not freeze fruits and vegetables in containers with a capacity over one-half gallon. Foods in larger containers freeze too slowly to result in a satisfactory product.

Packaging materials must have certain characteristics:

- Resistant to moisture and vapors.
- Durable, leakproof, easy to seal and mark
- Not become brittle and crack at low temperatures
- Resistant to oil, grease or water
- Protective against absorption of off-flavours or odours
- There are two types of packaging materials for home use:
  - Rigid containers and
  - Flexible bags or wrappings.

Rigid Containers

These are made of plastic or glass, are suitable for all packs and especially good for liquid packs. Straight sides on rigid containers make the frozen food much easier to get out. Rigid containers are often reuseable and make the stacking of foods in the freezer easier.

Cardboard cartons from cottage cheese, ice cream and milk are not sufficiently moisture and vapour resistant to be suitable for long-term freezer storage, unless they are lined with a freezer bag or wrap.

Regular glass peratures. The wide mouth allows easy removal of partially thawed foods. If standard canning jars are used for freezing, leave extra headspace to allow for expansion of foods during freezing. Expansion of the liquid could cause the jars to break at the neck. Some foods will need to be thawed completely before removal from the jar. Covers for rigid containers should fit tightly. If they do not, reinforce the seal with freezer tape. Freezer tape is especially designed to stick at freezing temperatures.

Flexible Bags or Wrappings

Bags and sheets of moisture-and vapour-resistant materials, laminated freezer papers, and heavy duty aluminum foil are suitable for dry-packed vegetables and fruits, meats, fish, or poultry. Bags can also be used for liquid packs. Protective cardboard cartons may be used to protect bags and sheets against tearing and to make stacking easier. Laminated papers are also used as protective over wraps. Plastic freezer bags are available in a variety of sizes. There are two types of closures. One type is twisted at the top, folded over and wrapped with twist ties included in the package. The other is zipped or pressed to seal a plastic channel. Regardless of type, press to remove as much air as possible before closing.

Vacuum Packaging

Vacuum packaging also works best for dry packs. Vacuum packaging removes more air from the packages than can be pressed out and seals the air out. Vacuuming packaging can help prevent oxidation which can lead to off-flavours and can save space in the freezer. Follow the directions that come with the vacuum packages carefully. To avoid the risk of botulism from some vacuum-packaged foods, keep foods frozen until ready to use.

Choose the Right Canner for Home Canning

Many myths exist about safe ways to can foods. The reality is that food must be processed using the right equipment and the right recipes to ensure safety. The type of canner you need depends on the food you want to preserve. A water bath canner is used to preserve acid foods like fruits and pickled foods. Low-acid foods, like vegetables and meats, must be canned in a pressure canner. Tomatoes can be processed in either type of canner, but always add lemon juice to be sure they are acid enough.

Boiling Water Bath Canner

A boiling water bath canner is simply a big pot with a removable rack to hold the jars, and a fitted lid. It must be big enough to cover the jars with 1 to 2 inches of boiling water. If you have an electric stove, be sure that the canner has a flat bottom, and that the canner is no more than 4 inches wider in diameter than the element or burner that you will be using. When the canner is centered on the element, the canner should not extend more than 2 inches on any side. This is important so that all the jars will be processed uniformly.

Pressure Canner: Not Pressure Cooker

Use a pressure canner not a pressure cooker. A pressure canner must be able to hold at least 4 quart-size jars. Most 16-quart or larger canners are big enough. All pressure canners have a removable rack, an automatic vent/cover lock, a vent port and a safety fuse. Only use canners that show the Underwriters Laboratories approval of safety symbol.

A pressure cooker or pressure saucepan is smaller and is not intended for processing foods in jars. A pressure cooker will heat up and cool down faster than a pressure canner, and the canned foods may not be processed long enough to destroy the microorganisms. Differences in the sizes of the pressure cookers and in the thickness of the food being processed make it impossible to adjust the times safely.

Dial-gauge or Weighted-gauge Canners

Dial-gauge canners have a dial that shows the pressure. These canners must be watched during processing to be sure the pressure does not fall below the required level and they must be tested every year to make sure the dial is registering correctly.

A dial-gauge canner is a good choice for those that live at higher altitudes because precise pressure adjustments are easy to make. Weighted-gauge canners have weights for 5, 10 or 15 pounds pressure that are placed over the vent and rock or jiggle when the correct pressure is reached.

Follow the manufacturer's instructions on how often the gauge should rock or jiggle. The sound of the weight rocking or jiggling indicates that the correct pressure is being maintained, so these canners do not need to be watched during processing. Weighted gauges do not have to be checked each year. A disadvantage for those living at higher altitudes is that precise pressure adjustments cannot be made as you are limited to 5, 10 or 15 pound weights.

Check your Pressure Canner for Safety

Whether your pressure canner is brand new, a family heirloom or a yard sale find, be sure all parts are in good condition:

- Rubber gaskets must be flexible and soft, not brittle, sticky or cracked. Check manufacturer's instructions for proper care of your gasket.
- Vent ports and openings must be clean and not clogged.
- The lid must not be warped and must fit properly and lock into place.
- Have the dial on your dial-gauge canner tested each year for accuracy.

Do a test run on your canner if it is new, or at the beginning of the season:

- Put several inches of water in your canner.
- Secure the lid and seal the vent.
- Turn on the heat and make sure the canner will get to the needed pressure and maintain it without leaking.
- Practice the correct way to depressurize the canner and remove the lid.

Canning Foods: the pH Factor

What does pH have to do with Canning Foods

The acidity, or pH, of foods determines how they must be processed for canning. Acid foods such as fruits and pickles with a pH of 4.6 or lower may be canned in a water bath canner.

Low-acid foods such as vegetables and meats with a pH above 4.6 must be processed in a pressure canner. Clostridium botulinum bacteria are the main reason why low-acid foods must be pressure canned to be safe. Clostridium botulinum spores can survive boiling water and grow in a sealed jar of low-acid food.

The spores can change into the vegetative cells that produce the deadly botulinum toxin. You must use a pressure canner to raise the temperature to the desired 240–250 °F to destroy the spores during the canning of low-acid foods. Some foods, such as figs and tomatoes, may be processed as acid foods, but because they may have pH values slightly above 4.6, lemon juice or citric acid must be added before canning.

Use a Tested Recipe

You do not need to know the pH of a food, but you must use a tested canning recipe based on the pH value of a food and other factors. In addition to pH, factors such as heat transfer properties, food particle size, food viscosity, and container size, influence the choice of an appropriate canning process.

Approximate pH of Some Foods

Considerable variation exists between varieties, condition of growing and processing methods. The pH values shown are for the edible portion of foods in their normal and natural state, unless indicated otherwise.

Fruits and Vegetables—with pH values near or above 4.6—must add acid according to tested recipes to bring pH to 4.6 or below:

- Figs, canned 4.92–5.00
- Papaya 5.20–6.00
- Tomatoes 4.30–4.90

Low-Acid Fruits—must pickle to lower the pH and process in a water bath canner:

- Cantaloupe 6.13-6.58
- Watermelon 5.18–5.60

Acid Fruits—may process in a water bath canner:

- Apple 3.30–4.00
- Blackberries, Washington 3.85–4.50
- Blueberries, Maine 3.12–3.33
- Cherries, California 4.01–4.54
- Gooseberries 2.80–3.10
- Muscadine grapes 3.20–3.40
- Nectarines 3.92–4.18
- Oranges, Florida 3.69–4.34
- Peaches 3.30–4.05
- Pears, Bartlett 3.50–4.60
- Pineapple 3.20–4.00
- Plums, Damson 2.90–3.10
- Plums, Red 3.60–4.30
- Raspberries 3.22–3.95
- Strawberries 3.00–3.90

Meat, Poultry and Seafood—must process in a pressure canner:

- Beef (ground) 5.1–6.2
- Chicken 6.2–6.4
- Clams 6.0–7.1
- Codfish, boiled 5.3–6.1
- Crab meat 6.5–7.0
- Ham 5.9–6.1
- Shrimp 6.5–7.0
- Veal 6.0
- Oysters 5.68–6.17

Low-Acid Vegetables—must process in a pressure canner or pickle to lower the pH:

- Artichokes, French (cooked) 5.60–6.00
- Artichokes, Jerusalem (cooked) 5.93–6.00
- Asparagus 6.00–6.70

- Beans, Lima 6.50
- Beans, String 5.60
- Beans, pork and tomato sauce (canned) 5.10–5.80

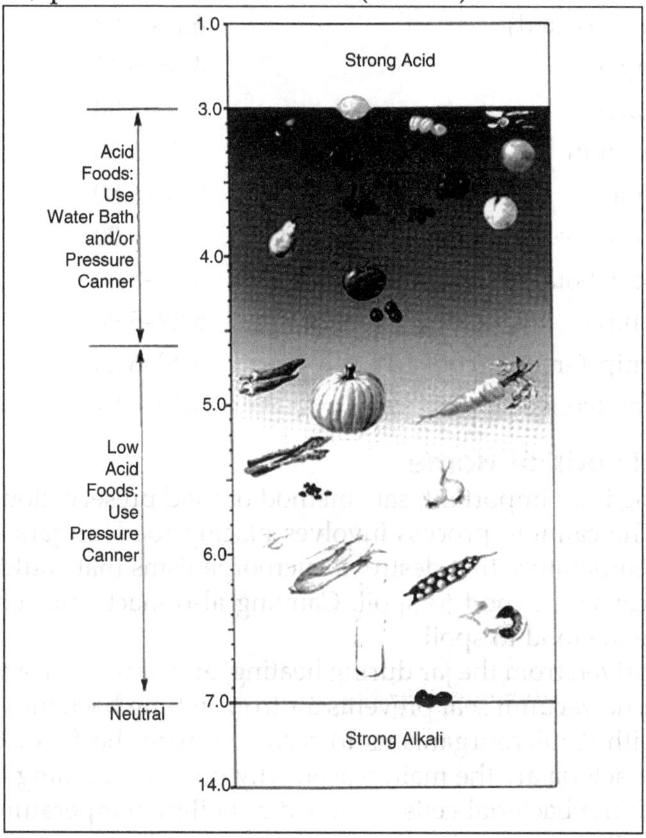

- Beets 5.30–6.60
- Broccoli, cooked 6.30–5.52
- Brussels sprouts 6.00–6.30
- Cabbage 5.20–6.80
- Carrots 5.88–6.40
- Cauliflower 5.60
- Corn 5.90–7.30
- Cucumbers 5.12–5.78
- Eggplant 5.50–6.50
- Garlic 5.80
- Hominy, cooked 6.00–7.50
- Mushrooms 6.00–6.70

• Okra (cooked)	5.50–6.60
• Onions, yellow	5.32–5.60
• Peas, Chick, Garbanzo	6.48–6.80
• Peas (cooked)	6.22–6.88
• Peppers	4.65–5.45
• Potatoes	5.40–5.90
• Pumpkin	4.90–5.50
• Spinach	5.50–6.80
• Squash, yellow (cooked)	5.79–6.00
• Sweet Potatoes	5.30–5.60
• Turnips	5.29–5.90
• Turnip Greens (cooked)	5.40–6.20
• Yams (cooked)	5.50–6.81

Canning Foods at Home

Canning is an important, safe method of food preservation if practiced properly. The canning process involves placing foods in jars and heating them to a temperature that destroys microorganisms that could be a health hazard or cause the food to spoil. Canning also inactivates enzymes that could cause the food to spoil.

Air is driven from the jar during heating, and as it cools, a vacuum seal is formed. The vacuum seal prevents air from getting back into the product bringing with it microorganisms to recontaminate the food. Clostridium botulinum bacteria are the main reason why pressure canning is necessary.

Though the bacterial cells are killed at boiling temperatures, they can form spores that can survive these temperatures. The spores grow well in low-acid foods, in the absence of air, such as in canned vegetables and meats.

When the spores begin to grow, they produce the deadly botulinum toxin. These spores can be destroyed by canning the food at a temperature of 240 °F, or above, for the correct length of time. This temperature is above the boiling point of water so it can only be reached in a pressure canner.

Safe Canning Methods

There are two safe ways of canning, depending on the type of food being canned. These are the boiling water bath method and the pressure canner method.

Boiling Water Bath Method

The boiling water bath method is safe for fruits, tomatoes and pickles as well as jams, jellies and other preserves. In this method, jars of food are heated by being completely covered with boiling water. High-acid foods contain enough acid that the Clostridium botulinum spores can't grow and produce their deadly toxin. High-acid foods include fruits and properly pickled vegetables. These foods can be safely canned at boiling temperatures in a boiling water bath. Tomatoes and figs have pH values close to 4.6. To can these in a boiling water bath, acid in the form of lemon juice or citric acid must be added.

Pressure Canning Methods

Pressure canning is the only safe method of canning low-acid foods. These include all vegetables, meats, poultry and seafood. Because of the danger of botulism, these foods must be canned in a pressure canner. Jars of food are placed in 2 to 3 inches of water in a pressure canner and then heated to a temperature of at least 240 °F. This temperature can only be reached in a pressure canner.

Getting Ready to can

Assemble and wash equipment and containers before gathering fruits and vegetables. Gather products early when they are at their peak of quality. Do not use overripe products.

Gather or purchase only as much as you can handle within 2 or 3 hours. Wash the product carefully, handling small amounts at a time.

Lift the food out of the water, drain the water and continue rinsing until the water is clear and free of dirt. Dirt contains some of the bacteria that are hardest to kill. Don't let the food soak; it will lose flavour and nutrients. The cleaner the raw foods, the more effective the canning process. Do not can decayed or damaged food items.

Preparing the Jars and Lids

Examine jars and discard those with nicks, cracks and rough edges. These defects will not permit an airtight seal on the jar, and food spoilage will result. All canning jars should be washed in soapy water, rinsed well and then kept hot. This could be done in the dishwasher or by placing the jars in the water that is heating in your canner.

The jars need to be kept hot to prevent breakage when they're filled with a hot product and placed in the canner for processing. Jars that will be filled with food and processed for less than 10 minutes in a boiling water bath canner need to be sterilized by boiling them for 10 minutes.

Note:

- If you are at an altitude of 1,000 feet or more, boil an additional minute for each 1,000 feet of additional altitude. Jars processed in a boiling water bath canner for 10 minutes or more or in a pressure canner will be sterilized during processing. Be sure to use new two-piece lids. Follow the manufacturer's instructions for treating them. Some need to be brought almost to a boil and then left in hot water, while others need to be boiled for a period of time.

Methods of Pack

Fruits and vegetables may be packed raw or they may be preheated and then packed into canning jars. The hot pack yields better colour and flavour, especially when foods are processed in a boiling water bath.

For both raw pack and hot pack, there should be enough syrup, water or juice to fill in around the solid food in the jar and to cover the food.

If not covered by liquid, food at the top tends to darken and develop unnatural flavours. It takes from ½ to 1½ cups of liquid for a quart jar.

Raw Pack

For this method put raw, unheated food directly in jars. Pour boiling hot water, juice or syrup over the food to obtain proper headspace.

Fruits and most vegetables packed raw should be packed tightly because they will shrink during processing; however, corn, lima beans, potatoes and peas should be packed loosely because they expand during canning.

Hot Pack

For this method, heat the food to boiling and then pack the hot food and boiling hot liquid in jars. Shrinkage has already taken place, so pack foods loosely enough to allow food to be surrounded by liquid.

Steps for Boiling Water Bath Method

- Use the boiling water bath method only to preserve high-acid foods such as fruits, tomatoes and pickles.
- Fill the canner about halfway with hot water. Turn on the burner and heat the water.
- For raw-packed jars, have the water in the canner hot, but not boiling, to prevent breakage of the jars when they're placed in the canner. For hot-packed jars, use hot or gently boiling water.
- Fill a single jar at a time, allow proper headspace, remove air bubbles, and adjust lids, then place the filled jar in the canner.
- Allow the proper headspace, according to processing directions for specific foods, in order to remove all the extra air during processing, and to form a tight vacuum seal.

- Run a bubble freer or any plastic or rubber-like utensil around the edges of the jar, gently shifting the food, so that any trapped air is released. Add more liquid as needed to to the jar to ensure proper headspace.
- Wipe off the rims of the jars with a clean, damp cloth.
- Screw on the lids, but not too tightly—air needs to escape during processing.
- Put filled glass jars on the rack in the canner. Add more boiling water or take out some as needed so that the water is at least 1 inch over the tops of the jars. If you add more water, pour it between the jars, not directly on them, to prevent breakage. Put the lid on the canner and set the heat to its highest setting.
- When the water in the canner reaches a rolling boil, begin counting the correct processing time. Adjust the heat to maintain a gentle and steady boil, and add more boiling water as necessary.
- If the water stops boiling at any time during the process, turn the heat on its highest setting, bring the water back to a vigourous boil, and begin the timing of the process over, from the beginning.
- When the jars have been processed in boiling water for the recommended time, turn off the heat and remove the canner lid. Wait 5 minutes before removing jars.
- Use a jar lifter to carefully remove the jars. Place the hot jars right side up on a cake cooling rack or dry towels to prevent the jars from breaking on contact with a cold surface.
- Avoid placing the jars in a cold draft. Do not cover with or wrap jars in towels. Leave at least 1 inch of space between jars.
- Allow the jars to cool untouched for 12 to 24 hours. Do not tighten the ring bands on the lids and do not push down on the center of the flat metal lids until the jars are completely cooled.

Steps for Pressure Canner Method

- Remember that pressure canning is the only safe method of processing low–acid foods such as vegetables, meat, poultry and fish. Be sure to read your manufacturer's instructions on the use of your pressure canner. Dial-gauge canners must be tested for accuracy every year before the canning season. Call your local Extension service office to make arrange-ments to have your dial-gauge canner tested.
- Place 2 to 3 inches of water in the canner. It should be hot, but not

boiling, when canning raw-packed food; hot or gently boiling for hot-packed foods. Make sure you still have 2 to 3 inches of water when you are ready to load the canner.

- Fill the jars as described in "Methods of Pack". Allow proper headspace, remove air bubbles, wipe jar rims and put on lids.
- Set the jars of food on the rack in the canner so steam can flow around each jar. Fasten the canner lid so that no steam begins to escape except through the vent. Turn heat to high and watch until steam begins to escape from the vent. Let the steam escape steadily for 10 minutes. This step is necessary for all pressure canners to remove air that could otherwise lower the temperature and result in underprocessing.
- Close the vent using a weight, valve or screw, depending on the type of canner. If you have a weighted-gauge canner that has a weight of varying pressures, be sure you are using the correct pressure.
- For a dial-gauge canner, let the pressure rise quickly to 8 pounds of pressure. Adjust the burner temperature down slightly and let the pressure continue to rise to the correct pressure. Start counting the processing time as soon as the pressure is reached.
- For weighted-gauge canners, let the canner heat quickly at first and then adjust the heat down slightly until the weight begins to rock gently or "jiggle" two to three times per minute, depending on the type of canner you have. Start counting the processing time as soon as the weight does either of these.
- Adjust the heat under the canner to maintain a steady pressure at the correct gauge pressure. If the pressure goes too high, turn down the heat under the canner. Do not lower the pressure by opening the vent or lifting the weight.
- Loss of pressure at any time can result in under processing or unsafe food. If at any time the pressure goes below the recommended amount, bring the canner back to pressure and begin the timing of the process over, from the beginning.
- When the processing is completed, carefully remove the canner from the heat. If the canner is too heavy, simply turn it off.
- Let the pressure in the canner drop to zero. This will take 30 to 45 minutes in a standard heavy-walled canner and nearly an hour for a larger 22-quart canner. Newer thin-walled canners depressurize more quickly. Do not rush the cooling by setting the canner in water or by running cold water over the canner. Never

lift the weight or open the vent to hasten the reduction in pressure. Forced cooling may result in food spoilage.

- Older canners are depressurized when the gauge on a dial-gauge canner registers zero or when a gentle nudge to the weight on a weighted gauge canner does not produce steam or resistance. New canners are equipped with a safety lock. These canners are depressurized when the safety lock drops to normal position. When the canner is depressurized, open the vent or remove the weight. Wait ten minutes and then open the canner.
- Sometimes safety locks that are located in the handle of a canner will stick. If a nudge to a canner weight shows that it is depressurized, remove the weight, wait 10 minutes and then run a knife blade between the handles to release the lock.
- Unfasten the lid, and tilt the far side up, so the steam escapes away from you. Do not leave the jars in the closed canner to cool, or the food inside could begin to spoil.
- Use a jar lifter to carefully remove the jars from the canner. Place the hot jars on a cake cooling rack or dry towels, right side up to prevent the jars from breaking on contact with a cold surface. Leave at least 1 inch of space between the jars.
- Do not tighten the lids. Allow the jars to cool, untouc-hed for 12 to 24 hours.

Safe Handling of Processed Jars

Test the Lid for a Proper Seal

Most two-piece lids will seal with a "pop" sound while they're cooling. When completely cool, test the lid. It should be curved downward and should not move when pressed with a finger. If a jar is not sealed, refrigerate it and use the unspoiled food within 2 to 3 days, reprocess within 24 hours, or freeze. If liquid has been lost from sealed jars do not open them to replace it, simply plan to use these first. The food may discolour, but if sealed, and the liquid is only a little lower than the food, the food is safe.

Label and Store Jars

Remove screw bands from the sealed jars to prevent them from rusting on. Wash, dry and store screw bands for later use. Wash food residue from the outside of the jars and rinse. Label, showing contents, date and lot number. Store in a clean, cool, dark, dry place. The best temperature is between 50 and 70°F. Avoid storing canned foods in a warm place near hot pipes, a range or a furnace, or in direct sunlight. They lose quality in a few weeks or months, depending on the temperature, and may even spoil. Keep canned goods dry. Dampness may corrode metal lids and cause leakage so food will spoil. For best quality, use canned foods within one year.

To Reprocess Food from Unsealed Jars

If you decide to reprocess food from jars that did not seal, do so within 24 hours. To do this, remove the lid and check the sealing surface on the jar for tiny nicks. Change the jar if necessary, add a new treated lid and reprocess using the same processing time. Label the jars of food that have been recanned and use these foods first. They will be softer in texture and lower in nutritional value than food processed only once.

How to Detoxify Canned, Low-Acid Foods

Contact with botulinum toxin can be fatal whether it is ingested or enters through the skin. Be extremely careful not to splash or come in contact with the suspect food or liquid. Wear disposable rubber or heavy plastic gloves. Wear clothes and aprons that can be bleached or thrown out if contaminated.

Step-by-Step Instructions for Detoxification:

- Carefully place the jars, with their lids, on their sides in an 8-quart, or larger pot or canner.
- Wash your gloved hands thoroughly.
- Carefully, without splashing, add enough hot water to the pot to completely cover the jars with at least 1 inch of water above the containers.
- Place a lid on the pot and heat the water to boiling.
- Boil for 30 minutes to make sure the food and containers are detoxified.
- Cool and discard the containers, their lids and food in the trash or dispose in a nearby landfill.

How to clean up contaminated surfaces:

- Wear rubber or heavy plastic gloves to clean up contaminated work surfaces and equipment, including can openers and clothing that may have come in contact with suspect foods or liquids.
- Use a fresh solution of 1 part unscented, liquid, house-hold, chlorine bleach to 5 parts clean water.
- Spray or wet contaminated surfaces with the bleach solution and let stand for 30 minutes. Avoid inhaling bleach or contact with skin.
- Wipe treated spills with paper towels and place paper towels in a plastic bag before putting them in the trash.

- Apply the bleach solution to all surfaces and equipment again, and let stand for 30 minutes and rinse.
- Wash all detoxified counters, containers, equipment, clothing, etc.
- Discard gloves when cleaning process is complete.

Testing for Sealed Jars

After 12 to 24 hours, test to be sure the jars are sealed. Most two-piece lids will seal with a "pop" sound while they're cooling.

When completely cool, test the lid. It should be curved downward and should not move when pressed with a finger. A conventional method is to tap the center of the lid with a spoon. A clear ringing sound means a good seal. A dull note may mean it doesn't have a tight seal or that food is touching the underside of the lid.

To determine which, hold the jar up and look at it. If no food is touching the lid, the jar does not have a tight seal. If a jar is not sealed, refrigerate it and use the unspoiled food within two to three days. Other options are to reprocess the food within 24 hours or to freeze it.

Freezing Basics

Freezing is one of the simplest and least time-consuming ways to preserve foods at home. Freezing does not sterilize food; the extreme cold simply retards growth of microorganisms and slows down changes that affect quality or cause spoilage in foods.

Freezing cannot improve the flavour or texture of any food, but when properly done it can preserve most of the quality of the fresh product. Select only the best-quality fruits and vegetables at their peak of maturity for freezing

Packaging and Labeling Tips

- Quickly cool all foods and syrup before packaging. This speeds up freezing and helps retain the natural colour, flavour and texture of food.
- Pack foods in quantities that will be used for a single meal. Do not freeze fruits and vegetables in containers with a capacity over one-half gallon. Foods in larger containers freeze too slowly to result in a satisfactory product.
- Follow directions for individual foods to determine which can be packed dry and which need added liquid. Some loose foods such as blueberries may be "tray packed".
- Pack food tightly, leaving as little air as possible in the package.
- Most foods require headspace between the packed food and closure

to allow for expansion of the food as it freezes. Foods that are exceptions and do not need headspace include loose packing vegetables such as asparagus and broccoli, bony pieces of meat, tray-packed foods and bread.

- Seal rigid containers carefully. Use a tight lid and keep the sealing edge free from moisture or food to ensure a good closure. Secure loose-fitting covers with freezer tape.
- Meats may be packaged using either the drugstore wrap or the butcher wrap. Label each package, including the name of the product, any added ingredients, packaging date, the number of servings or amount, and the form of the food, such as whole, sliced, etc. Use freezer tape, marking pens or crayons, or gummed labels that are made specifically for freezer use.

Drugstore Wrap

1. Center meat on paper.

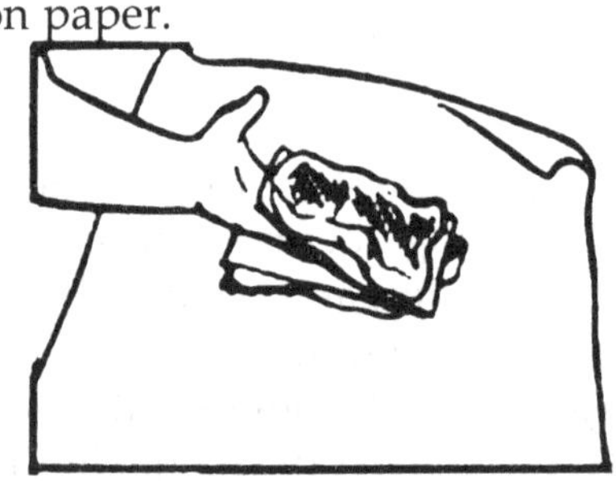

2. Bring two sides of paper together at top.

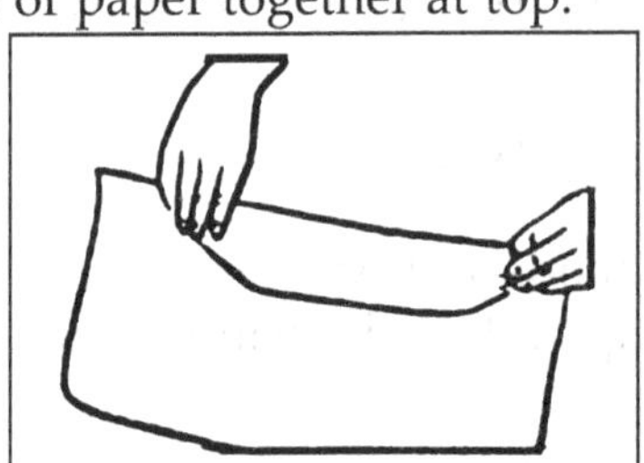

3. Fold down about ½ to ¾ inch.

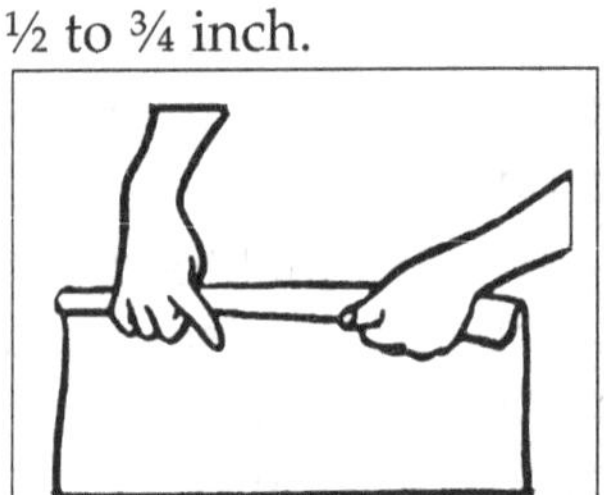

4. Roll folded edge down until snug against meat.

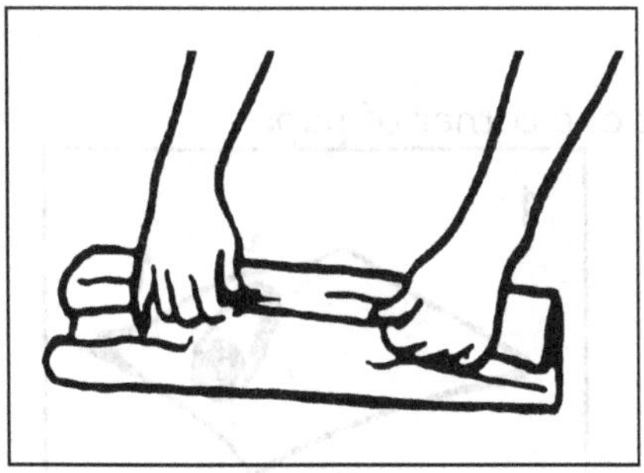

5. Turn package over. Press out air from sides.

6. Fold ends into triangles.

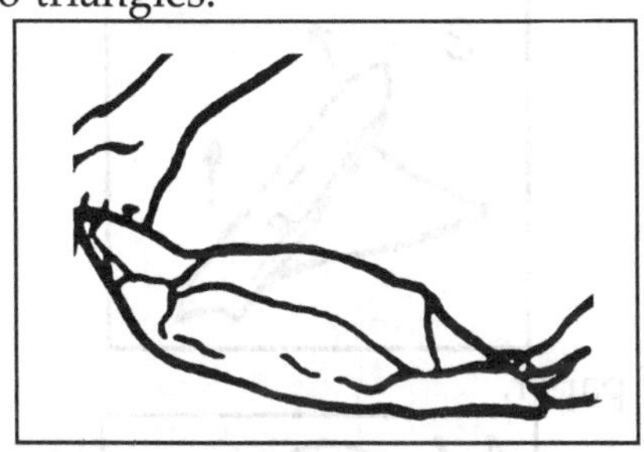

7. Bring to center and tape to secure.

8. Label and date.

Butcher Wrap

1. Place meat at one corner of paper.

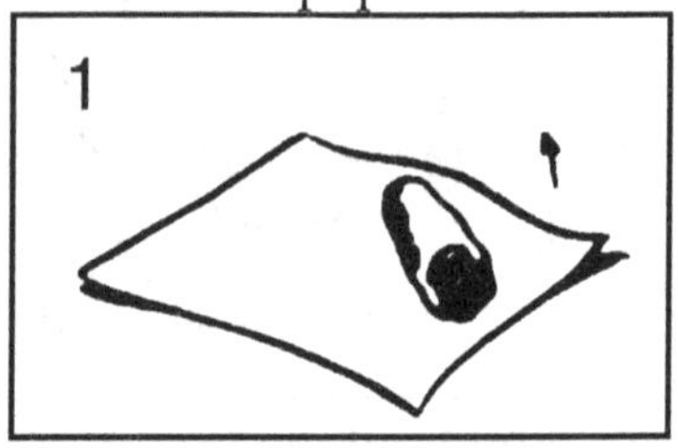

2. Roll up tightly towards opposite corner.

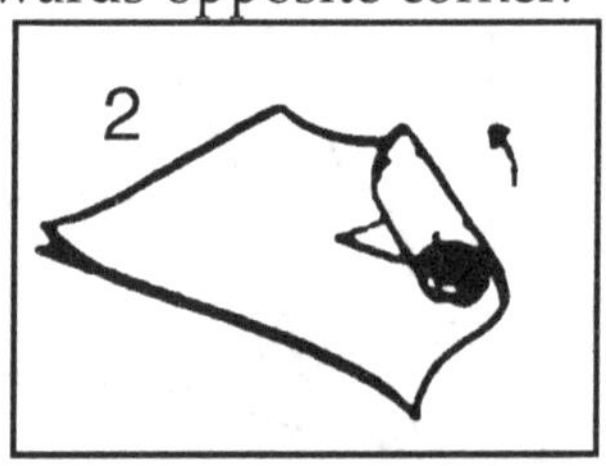

3. Tuck sides in.

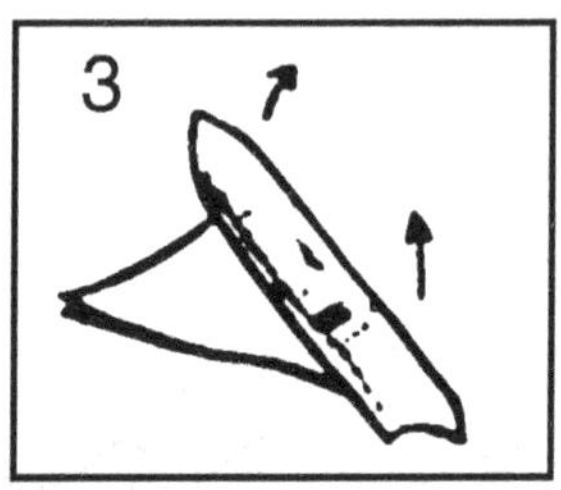

4. Roll to end of paper.

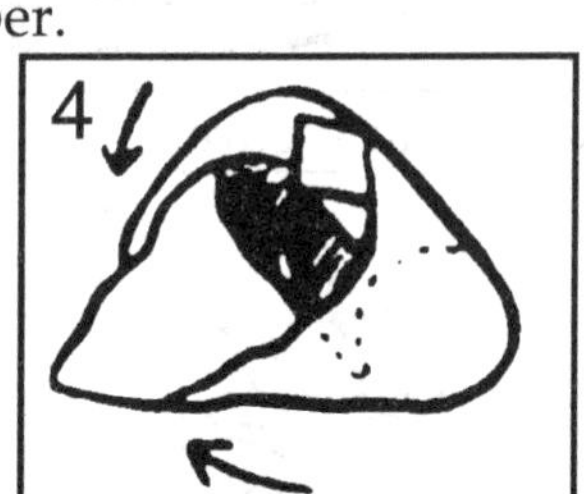

5. Seal open edges with freezer tape.

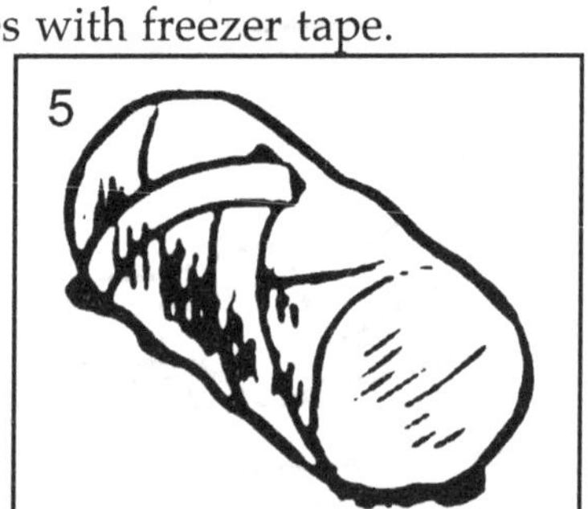

Table. Head Space to Allow between Packed Food and Closure.

Type of Pack	Container with Wide Top Opening		Container with Narrow Top Opening	
	Pint	Quart	Pint	Quart
Liquid pack*	½ inch	1 inch	¾ inch***	1½ inch
Dry pack**	½ inch	½ inch	½ inch	½ inch

Notes:

*Fruit packed in juice, syrup or water; crushed or puréed fruit; or fruit juice.

**Fruit or vegetables packed without added sugar or liquid. ***Headspace for juice should be 1½ inches.

Freezer Pointers

- Freeze food at 0°F or lower. To facilitate more rapid freezing, set the temperature control at minus 10 °F or lower about 24 hours in advance.
- Freeze foods as soon as they are packaged and sealed.
- Do not overload your freezer with unfrozen food. Add only the amount that will freeze within 24 hours; usually 2 to 3 pounds of food per cubic foot of storage space. Overload slows down the freezing rate, and foods that freeze too slowly may lose quality.
- Place packages in contact with refrigerated surfaces in the coldest part of the freezer.
- Leave a little space between packages so air can circulate freely. Then, when the food is frozen, store the packages close together.

Freezer Management

A full freezer is most energy efficient. Post a frozen foods inventory near the freezer and keep it up-to-date by listing the foods and dates of freezing as you put them in the freezer. Check them off as you take them out. By keeping an inventory, you will know the exact amounts and kinds of foods that are in the freezer at all times. It also helps to keep foods from being forgotten. Maintain the storage temperature at 0 °F or lower. At higher temperatures, foods lose quality much faster. Keep a freezer thermometer in your freezer and check the temperature frequently.

Foods that do not Freeze Well

- Cabbage, celery, cress, cucumbers, endives, lettuce, parsley, radishes
- Cheese or crumb toppings

- Cream or custard fillings, milk sauces, sour cream
- Cooked egg whites, icings made from egg whites
- Fried foods
- Fruit jelly, gelatin
- Irish potatoes, baked or boiled
- Cooked macaroni, spaghetti or rice

Effect of Freezing on Spices and Seasonings

When using seasonings and spices, season lightly before freezing, and add additional seasonings when reheating or serving.

Pepper, cloves, garlic, celery seasonings, green pepper, imitation vanilla and some herbs tend to get stronger and bitter. Onion, paprika and curry change flavour during freezing. Salt loses flavour and has the tendency to increase rancidity of any item containing fat.

Table. Remedies for Freezing Problems.

Problems	Cause	Prevention
Freezer burn-(Surface of food light colored and dried out; food is tough, dry and less flavorful)	Torn or unsealed packages	Be sure all packages are tightly sealed so no air can get in. Handle carefully to avoid tears.
	Packaging not moisture- and vapor-resistant	Use only packaging approved for use in freezing.
	Too much air in package	Always press out all air in wrapped foods. Use just the right size containers for the amount of food and proper headspace.
Gummy liquid in fruits	Fruits frozen too slowly or freezer temperatures too warm or fluctuating	Freeze foods at 0 °F or below immediately after packaging and maintain that temperature throughout storage. Do not freeze more than 2 to 3 pounds of food per cubic foot of storage space at one time.
Mushy food	Large ice crystals form in the food, breaking the cell structure	
"Grassy" flavors in vegetables	Freezing unblanched vegetables	Blanch all vegetables as directed.
Green vegetables turn olive brown colors		
Rancid flavors - (strong oily-like flavor in fats)	Spoilage of the fat in a product	Blanch all vegetables as directed. Package food correctly and exclude all unnecessary air. Do not store for longer than recommended.
Maroon-colored bones or pink meat in frozen poultry after it is cooked	Hemoglobin (coloring matter) in bones	Natural occurrence. There's nothing wrong with the meat.

Freezing Fruits and Vegetables

Freezing is a quick and convenient way to preserve fruits and vegetables at home. It is the method of food preservation that preserves the greatest quantity of nutrients.

To maintain top nutritional quality in frozen fruits and vegetables, it is essential to:

- Select fresh, firm-ripe produce
- Blanch vegetables as directed
- Store the frozen product at 0 °F
- Use within suggested storage times.

Chemical Changes

Harvested fresh, fruits and vegetables continue to undergo chemical changes that can cause spoilage and deterioration of the product.

This is why these products should be frozen as soon after harvest as possible and at their peak degree of ripeness. Enzymes in the fruits and vegetables must be inactivated to prevent the loss of nutrients and colour and flavour changes that will occur.

Enzymes in Vegetables

Enzymes are inactivated by the blanching process, which is the exposure of the vegetables to boiling water or steam for a brief period of time. The vegetable must then be cooled rapidly in ice water to prevent it from cooking. Contrary to statements in some publications on home freezing, in most cases blanching is absolutely essential for producing top-quality frozen vegetables. Blanching also helps to destroy microorganisms on the surface of the vegetable, brightens the colour, helps retard loss of vitamins and helps make some vegetables, such as broccoli and spinach, more compact.

Enzymes in Fruits

The major problems associated with enzymes in fruits are the development of brown colours and loss of vitamin C. Because fruits are usually served raw, they are not blanched like vege-tables. Instead, ascorbic acid is used either in its pure form or in commercial mixtures, such as Fruit Fresh™, to control the activity of the enzymes. Other methods to control browning include soaking the fruit in dilute vinegar solutions or coating the fruit with sugar and lemon juice. However, these other methods do not prevent browning as effectively as treatment with ascorbic acid.

Preventing Rancid Flavours

Rancid oxidative flavours may develop through contact of the frozen product with air. This problem can be prevented by using a wrapping material that does not permit air to pass into the product and by removing as much air as possible from the freezer bag or container before freezing.

Textural Changes

Cause of Textural Changes

Water makes up over 90 per cent of the weight of most produce and is held within the fairly rigid cell walls that give support, structure and texture to the fruit or vegetable. Freezing fruits and vegetables actually consists of freezing the water contained in the plant cell.

When the water freezes, it expands, and the ice crystals cause the cell walls to rupture.

The texture of the thawed produce will be much softer than it was when raw. For example, when a frozen tomato is thawed, it turns into a pile of mush and liquid. This explains why celery, lettuce and tomatoes are not usually frozen, and why it is recommended that raw frozen fruits be eaten before they have completely thawed. Textural changes due to freezing are not as apparent in products that are cooked before eating because cooking also softens cell walls. These changes are also less noticeable in high-starch vegetables, such as peas, corn and lima beans.

Prevention

When produce is frozen quickly, it forms a large number of small ice crystals. These small ice crystals produce less cell wall rupture than slow freezing, which produces only a few large ice crystals.

This is why it is sometimes recommended that the temperature of a freezer be set at the coldest setting several hours before foods will be placed in the freezer. The maximum amount of unfrozen product that can be frozen at one time is 2 to 3 pounds to each cubic foot of freezer space per 24 hours.

Moisture Loss

Moisture loss, or ice crystals evaporating from the surface area of a product, produces freezer burn—a grainy, brownish spot where the tissues become dry and tough. This surface freeze-dried area is very likely to develop "off" flavours. Packaging in heavyweight, moisture-proof wrap will prevent freezer burn.

Microbial Growth in the Freezer

The freezing process does not actually destroy the micro-organisms that may be present on fruits and vegetables. While blanching destroys some microorganisms and there is a gradual decline in the number during freezer storage, sufficient populations are still present to multiply and cause spoilage of the product when it thaws. For this reason it is necessary to inspect carefully any frozen products that have accidentally thawed if the power goes off or the freezer door is left open.

Recommended Storage Times

Longer storage of fruits and vegetables than those recommended will not make the food unsafe for use but will decrease its quality.

Fruits

Most frozen fruits maintain high quality for eight to 12 months. Unsweetened fruits lose quality faster than fruits packed in sugar or sugar syrups.

Vegetables

Most vegetables will maintain high quality for 12 to 18 months at 0 °F or lower. However, it is a good idea to use your home-frozen vegetables before the next year's crop is ready for freezing.

Selecting Freezer Containers

Use good-quality freezer containers that are both moisture-and vapour-proof so that moisture can be kept in the product and air kept away from it.

Moisture–and Vapour–Resistant Wraps

Heavyweight aluminum foil, plastic-coated freezer paper and saran are effective at excluding oxygen. They should be strong, pliable and adhere to the shape of the food item. Seal these only with tape that is designed for the freezer because other household tapes lose adhesive quality in the extremely cold freezer temperatures. Wraps are not as convenient for fruits and vegetables as plastic bags or rigid freezer containers.

Plastic Film Bags

These seal with twist-and-tie tops. Collapsible cardboard freezer boxes are frequently used as an outer covering for plastic bags to protect them against tearing and for easy stacking in the freezer. Plastic sandwiches bags and bread wrappers are not suitable for freezing.

"Freeze–and–Cook" Bags

These bags withstand temperatures from below 0 °F to above the boiling point and are suitable for both freezing and cooking the product. They come in 1½ pint and quart sizes and also as large rolls of plastic so they can be made the size desired. A heat sealer is necessary for closing these bags. These products are more expensive but convenient since you can cook in them.

Freezer Temperature

To maintain top quality, frozen fruits and vegetables should be stored at 0 °F or lower. Higher temperatures increase the rate at which deterioration can take place and can shorten the shelf life of frozen foods. Do not attempt to save energy by raising the temperature of frozen food storage above 0 °F. A freezer thermometer can help determine the actual temperature of the freezer. If the freezer has number temperature settings, such as from one to nine, check the manual to see what settings are recommended for different uses. Changing temperatures in the freezer can cause the migration of water vapors from the product to the surface of the container as may be seen in improperly handled commercially frozen foods.

Methods of Packing Fruits

There are three ways to pack fruits for freezing: sugar pack, syrup pack and unsweetened pack. Although some fruits may be packed without sweeteners, the flavour of many fruits is retained better with the use of sugar. Gooseberries, currants, cranberries, blueberries and rhubarb give as good quality packs with or without sugar.

Unsweetened Pack

Simply spread a single layer of prepared fruit on shallow trays and freeze. To prevent freezer burn, package the fruit as soon as it is frozen and return to the freezer. Sugar substitutes may be used in any of the unsweetened packs. Both saccharin and aspartame work well in frozen products, or they can be added to the fruit just before serving.

Sugar Pack

To freeze fruits using a sugar pack, sprinkle the required amount of sugar over the fruit. Gently stir until the pieces are coated with sugar and juice.

Syrup Pack

To make sugar syrup, dissolve the needed amount of sugar in water, mixing until the solution is clear. Chill syrup before using. Use just enough cold syrup to cover the fruit.

Preventing Discolouration in Fruits

Some fruits, such as peaches, apples, pears and apricots, darken quickly when exposed to air and during freezing. They may also lose flavour when thawed. There are several ways to prevent darkening of fruit and flavour loss.

Ascorbic Acid (Vitamin C)

Ascorbic acid or vitamin C is effective in preventing discolouration in most fruits. Not only does it preserve natural colour and flavour of fruits, but it adds nutritive value as well. Ascorbic acid in powdered form is available at some drugstores or where freezing supplies are sold. Ascorbic acid tablets may be more readily available and less expensive, but are more difficult to dissolve. They do need to be finely crushed before use. Fillers in the tablets may make the syrup cloudy, but they are not harmful. One-half teaspoon powdered ascorbic acid equals 1500 mg.

Ascorbic acid may be added by the following methods when freezing fruits:

- *Syrup or Liquid Packs:* Add powdered or crushed ascorbic acid to cold syrup shortly before using. Stir it gently so you do not stir in air. Keep syrup refrigerated until use.
- *Sugar or Dry Packs*: Dissolve the ascorbic acid in two to three tablespoons of cold water and sprinkle dissolved ascorbic acid over fruit just before adding sugar.
- *Crushed Fruits, Fruit Pulses and Fruit Juices:* Add ascorbic acid to prepared fruit and stir well.

Ascorbic Acid Mixture

Ascorbic acid mixtures are special anti-darkening preparations, usually made of ascorbic acid mixed with sugar or with sugar and citric acid.

The important active ingredient in these mixtures is ascorbic acid. Follow the manufacturer's directions for use.

Citric Acid and Lemon Juice

Citric acid or lemon juice are sometimes used in place of ascorbic acid but are not as effective as ascorbic acid. When used in large quantities, they often mask natural fruit flavours.

Steps in Freezing Fruits:

- Select best-quality fruits; wash and sort fruits carefully, discarding parts that are of poor quality. Do not soak.
- Prepare fruits as you will use them. Do not use iron or copper equipment that can react with the acid in fruit.
- Use ascorbic acid as an anti-browning treatment.
- Use dry sugar or sugar syrup as recommended. If you are preparing a sugarless pack of fruits that brown, be sure to treat with ascorbic acid or other antibrowning agents.
- Pack into good-quality plastic bags, freezer containers or freezer jars. Allow ½-inch headspace for expansion. Keep fruits that tend to darken, such as peaches, under the syrup by placing crumpled wax paper between lid and fruit.

To Use Home–Frozen Fruits

Thaw fruit at room temperature in its original package to preserve quality and nutritive value. If faster defrosting is required, place package in front of an electric fan or submerge in cool lukewarm water. Serve as soon as defrosted, preferably while a few ice crystals remain.

Methods of Packing Vegetables

There are two basic methods for packing vegetables for freezing, the tray pack and the dry pack.

Dry Pack

This is the term used to describe the packing of blanched and drained vegetables into containers or freezer bags. Pack the vegetables tightly to cut down on the amount of air in the container.

If the vegetables are packed in freezer bags, press air out of the unfilled part of the bag. When packing broccoli, alternate the heads and stems. Allow ½ inch headspace.

Tray Pack

This is the method of freezing individual pieces of blanched and drained vegetables on a tray or shallow pan, then packing the frozen pieces into a freezer bag or container. This method produces a product similar to commercially frozen plastic bags of individual vegetable pieces and is particularly good for peas, corn and beans. Pack the frozen pieces into a bag or container as soon as they are frozen. Long exposure will result in loss of moisture. Do not leave headspace. Seal.

Steps for Freezing Vegetables:

- Assemble the necessary equipment.
  - Large pot
  - A colander, wire basket, or net bag for blanching
  - Large pans for cooling
  - Ice cubes or ice blocks for cooling
  - Cutting board, knives, hot pads
  - Plastic freezer bags or other containers
  - A timer or a clock with a second hand
- *Choose Vegetables for Freezing that are at their Peak of Flavour and Texture*: If possible, harvest the vegetables in the cool of the morning and process immediately or refrigerate the vegetables until processed to preserve quality and nutrients.

- *Carefully Follow the Blanching Instructions*: Time the blanching from when the vegetable is immersed in the vigourously boiling water. Do not add so much vegetable that the water stops boiling. Under blanching stimulates the activity of enzymes and is worse than no blanching. Over blanching causes loss of flavour, colour, vitamins and minerals.

Methods of Blanching Vegetables

Boiling Water Blanching

- Use 1 gallon water for each pound of vegetables except for leafy greens, which need 2 gallons per pound.
- Bring water to rolling boil.
- Immerse wire basket, blanching basket or mesh bag containing vegetables.
- Cover pot and boil at top heat the required length of time. You may use the same blanching water two or three times. Keep it at the required level. Change the water if it becomes cloudy.
- Cool vegetables immediately in pans of ice water for the same time used for blanching. Keep water ice cold for chilling.
- Drain the vegetables thoroughly. Extra water will form too many ice crystals.
- Pack, using dry or tray pack methods. Freeze.

Water Blanching Times

- *1½ Minutes:* Cabbage (shredded)
- *2 Minutes*: Asparagus (small stalks), blackeye peas, carrots (diced or sliced), greens (except collards)
- *3 Minutes:* Beans (snap, green or wax), broccoli and cauliflower flowerets (1½ inches across), celery, collards, sweet peppers (halves), rutabagas and turnips (cubed), summer squash (½-inch slices)
- *4 Minutes:* Whole kernel or cream corn (blanch on cob, cool and cut off cob), eggplant (1/3-inch slices)
- *5 Minutes*: Carrots (whole, small)

The following times depend on size:

- *1½–2½ Minutes:* Green peas
- *2–4 Minutes:* Asparagus, beans: lima, butter, pinto
- *3–4 Minutes:* Okra
- *3–5 Minutes*: Brussel sprouts, Irish potatoes (new)

- *3–7 Minutes*: Onions (until center is heated)
- *7–11 Minutes*: Corn-on-the-cob

Steam Blanching

Blanching in steam is recommended only for a few vegetables.

For broccoli, pumpkin, sweet potatoes and winter squash, both steaming and boiling work well:

- Use a pot with a tight lid and a basket that holds the food at least 3 inches above the bottom of the pot.
- Put 1–2 inches of water in the pot and bring the water to boil.
- Put the vegetables in the basket in a single layer so that steam reaches all parts quickly.
- Cover the pot and keep heat high.
- Start counting steaming time as soon as the lid is on.

Steam blanch broccoli flowerets (1 ½ inches across) for 5 minutes. Steam the other recommended vegetables until soft before freezing.

Microwave Oven–Blanching

Directions for microwave-blanching of vegetables are not based on any published research. Recent research indicates some of these times are not sufficient to adequately blanch some vegetables. Inadequate blanching does not cause a food safety problem, but quality may suffer. Water is a more reliable method of blanching.

To Use Home–Frozen Vegetables

All vegetables may be cooked from the frozen state except corn-on-the-cob, which should be partially defrosted. Cook frozen vegetables in a small amount of salted water.

Cook only until tender—about half as long as if the vegetables were fresh. You can use a pressure saucepan or microwave oven for cooking frozen vegetables. Follow manufacturer's directions for cooking time. A pack should be thawed enough to break it up before pressure cooking. If cooking vegetables in a freezer pouch in the microwave oven, be sure to split the package open first.

Freezing Meats and Seafood

Safety of Frozen Meats

Wholesome food stored constantly at 0 °F will always be safe. Only the quality suffers with lengthy freezer storage. Freezing preserves food for extended periods because it prevents the growth of microorganisms that cause both food spoilage and food borne illness. Once thawed, however, these microbes can again become active, multiplying under the right conditions to levels that can lead to food borne illness. Since they will then grow at about the same rate as microorganisms on fresh food, handle thawed items as you would any perishable. Trichina and other parasites can be destroyed by sub-zero freezing temperatures. However, very strict government-supervised conditions must be met. It is not recommended to rely on home freezing to destroy trichina. Thorough cooking will destroy all parasites.

Nutrient Value

The freezing process itself does not destroy nutrients. In meat and poultry products, there is little change in nutrient value during freezer storage.

Packaging

Proper packaging helps maintain quality and prevent "freezer burn." It is safe to freeze meat or poultry directly in its supermarket wrapping, but this type of wrap is permeable to air.

Unless you will be using the food in a month or two, overwrap these packages as you would any food for long-term storage using airtight heavy-duty foil, plastic wrap or freezer paper, or place the package inside a plastic bag. Use these materials or airtight freezer containers to repackage family packs into smaller amounts or freeze foods from opened packages. It is not necessary to rinse meat and poultry before freezing. Freeze unopened vacuum packages as is. If you notice that a package has accidentally torn or has opened while food is in the freezer, it is still safe to use; merely overwrap or rewrap it.

Freezer Burn

Freezer burn does not make food unsafe, merely dry in spots. It appears as grayish-brown leathery spots and is caused by air reaching the surface of the food.

Cut freezer-burned portions away either before or after cooking the food. Heavily freezer-burned foods may have to be discarded for quality reasons.

Colour Changes

Colour changes can occur in frozen foods. The bright red colour of meat as purchased usually turns dark or pale brown depending on its variety. This may be due to lack of oxygen, freezer burn or abnormally long storage.

Freezing doesn't usually cause colour changes in poultry. However, the bones and the meat near them can become dark. Bone darkening results when pigment seeps through the porous bones of young poultry into the surrounding tissues when the poultry meat is frozen and thawed.

Freeze Rapidly

Freeze food as quickly as possible to maintain its quality. Slow freezing creates large, disruptive ice crystals. During thawing, they damage the cells and cause meat to "drip" or lose juiciness. Ideally, food 2 inches thick should freeze completely in about two hours. If your home freezer has a "quick-freeze" shelf, use it. Never stack packages to be frozen. Instead, spread them out in one layer on the shelves, stacking them only after frozen solid.

Refrigerator Freezers

If a refrigerator freezing compartment can't maintain 0 °F, or if the door is opened frequently, use it only for short-term food storage.

Eat those foods as soon as possible for best quality. Use a free-standing freezer set at 0 °F or below for long-term storage of frozen foods. Keep a thermometer in your freezing compartment or freezer to check the temperature.

Length of Time

Because freezing keeps food safe almost indefinitely, recommended storage times are for quality only. If a food is not listed on the chart, you may determine its quality after defrosting. First check the odour. Some foods will develop a rancid or off odour when frozen too long and should be discarded. Some may not look picture-perfect or be of high enough quality to serve alone, but may be edible; use them to make soups or stews. Cook raw food and if you like the taste and texture, use it.

Freezing Game Meats

Freshly slaughtered meat carcasses or pieces need to be cooled to below 40 °F within 24 hours to prevent souring or spoiling. The meat should be chilled to 32 to 36 °F. Variety meats are ready to be wrapped and frozen after they are cold.

Quail, dove, duck, pheasant and other game birds should be dressed and gutted as soon as possible after shooting. Cool and clean properly. Remove excess fat on wild ducks and geese since it becomes rancid very quickly. Freeze as directed for poultry.

Note

Do not stuff poultry or game birds before freezing them. During freezing or thawing, food poisoning bacteria could easily grow in the stuffing. Commercially-stuffed frozen poultry is prepared under special safety conditions that cannot be duplicated at home.

Freezing Fish

Fish for freezing should be fresh as possible. Wash fish and remove scales by scraping fish gently from tail to head with the dull edge of the knife or spoon.

Remove entrails after cutting entire length of belly from vent to head. Remove head by cutting above collarbone. Break backbone over edge of cutting board or table. Remove dorsal or large back fin by cutting flesh along each side and pulling fin out. Do not trim fins with shears or a knife because bones will be left at the base of the fin. Wash fish thoroughly in cold running water.

Fish is now dressed or pan dressed, depending on size. Large fish should be cut into steaks or fillets for easier cooking. For steaks, cut fish crosswise into ¾-inch thick steaks. For fillets, cut down back of fish from tail to head. Then cut down to backbone just above collarbone.

Turn knife flat and cut flesh along backbone to tail allowing knife to run over rib bones. Lift off entire side of fish in one piece, freeing fillet at tail. Turn fish over and cut fillet from other side.

Pretreating

Fish are categorized as either fat or lean fish by the amount of fat in their flesh. Fat fish include varieties such as mullet, mackerel, trout, tuna and salmon. Lean fish include flounder, cod, whiting, redfish, croaker, snapper, grouper, sheepshead and most freshwater fish.

Before freezing, fish can be pretreated to improve the quality of the stored fish. Fatty fish should be dipped for 20 seconds in an ascorbic acid solution made from 2 tablespoons ascorbic acid to 1 quart of cold water to control rancidity and flavour change.

Lean fish may be dipped for 20 seconds in a brine of ¼ cup salt to 1 quart of cold water to firm the fish and decrease drip loss on thawing.

Methods of Freezing

Fish may be frozen using any of the following glazes. If several fish are placed in the same package, place freezer paper or wrap between them for easier separation.

- *Lemon-Gelatin Glaze*: To prepare glaze, mix ¼ cup lemon juice and 1¾ cups water. Dissolve one packet of unflavoured gelatin in ½ cup lemon juice-water mixture. Heat the remaining 1½ cups of liquid to boiling. Stir the dissolved gelatin mixture into the boiling liquid. Cool to room temperature. When cool, dip the cold fish into the lemon-gelatin glaze and drain. Wrap the fish in moisture-and vapour-resistant packaging, label and freeze.

- *Ice Glaze*: Place unwrapped fish in the freezer to freeze. As soon as it is frozen, dip fish in near-freezing ice water.

  Place fish again in the freezer a few minutes to harden the glaze. Take fish out, and repeat the glaze until a uniform cover of ice is formed. Wrap the fish in moisture-and vapour-resistant paper or place in freezer bags. Label and freeze.
- *Water*: Place fish in a shallow metal, foil or plastic pan; cover with water and freeze. To prevent evaporation of the ice, wrap the container in freezer paper after it is frozen, label and freeze.

Fish Roe

Thoroughly wash and package in freezer containers or bags and boxes, leaving ¼-inch head space. Seal and freeze.

Freezing Shellfish

Clams

Clams can be frozen either in the shell or shucked. To freeze the clams in the shell, simply place the live clams in moisture-and vapour-resistant bags. Press out excess air and freeze. To freeze the clam meat, shuck the clams, then clean and wash the meat thoroughly. Drain and pack in freezer containers, leaving ½-inch head-space. Seal, label and freeze.

Crabs

Select only live crabs to prepare for freezing. Crab freezes better if not "picked" before freezing. Simply remove the back, legs, entrails and gills either before or after boiling the crab for five minutes.

The claws and body or core of the crab that still contains the meat should then be wrapped or ice-glazed and wrapped in freezer wrap or paper.

Lobster

For best quality, lobster should be frozen uncooked. Freeze the lobster whole, or clean it and freeze just the shell portions that contain the edible meat.

Freeze lobster in the shell, to help keep the meat from drying out. Simply wrap the whole lobster or lobster portions in moisture-and vapour-resistant wrapping and freeze. Lobster can be cooked and then frozen, but the quality will not be as good.

Oysters

Oysters that are still in the shell should only be frozen live. A live oyster will keep its shell tightly closed or will close when tapped.

If you have plenty of freezer space and want to freeze the oysters in the shell, simply wash the shells thoroughly and place in moisture-and vapour-resistant bags. Shuck the oysters to save freezer space.

First, wash the oyster shells, discarding any that have died. Shuck oysters into a strainer, saving the liquor, and remove any pieces of shell or sand. If necessary, the oysters can be rinsed to remove any sand. Place oysters and liquor in a plastic container or freezer bag, leaving ½-inch headspace. Seal and freeze.

Scallops

Scallops for freezing should be live until shucked. A live scallop will keep its shell tightly closed or will close it when tapped. To freeze, place shucked scallops in a freezer container, leaving ½-inch headspace, seal and freeze.

Shrimp

Select high-quality, fresh shrimp for freezing. Shrimp can be frozen cooked or raw, in or out of the shell. For maximum storage life and quality, freeze shrimp raw, with heads removed, but shells still on. Shrimp may also be frozen in water in a freezer container.

Be sure to wash and drain the shrimp if frozen uncooked. Quickly chill cooked shrimp before freezing. Package in freezer containers or bags, leaving ¼-inch headspace. Seal and freeze.

Safe Defrosting

Never defrost foods in a garage, basement, car, plastic garbage bag, out on the kitchen counter, outdoors or on the porch. These methods can leave your foods unsafe to eat. There are three safe ways to defrost food: in the refrigerator, in cold water or in the microwave.

It's best to plan ahead for slow, safe thawing in the refrigerator. Small items may defrost overnight; most foods require a day or two. For large items like turkeys allow one day for each 5 pounds of weight.

For faster defrosting, place food in a leak-proof plastic bag and immerse it in cold water. Check the water frequently to be sure it stays cold.

Change the water every 30 minutes. After thawing, refrigerate the food until ready to use. When microwave-defrosting food, plan to cook it immediately after thawing because some areas of the food may become warm and begin to cook during microwaving. Holding partially cooked food is not recommended because any bacteria present wouldn't have been destroyed.

Refreezing

Once food is thawed in the refrigerator, it is safe to refreeze it without cooking, although there may be a loss of quality due to the moisture lost through defrosting. After cooking raw foods that were previously frozen, it is safe to freeze the cooked foods. And if previously cooked foods are thawed in the refrigerator, you may refreeze the unused portion. If you purchase previously frozen meat, poultry or fish at a retail store, you can refreeze if it has been handled properly.

Cooking Frozen Foods

Raw or cooked meat, poultry or casseroles can be cooked or reheated from the frozen state. However, it will take approximately one and a half times the usual cooking time for food which has been thawed. Remember to discard any wrapping or absorbent paper from meat or poultry.

When cooking whole poultry, remove the giblet pack from the cavity as soon as you can loosen it. Cook the giblets separately. Read the label on USDA-inspected frozen meat and poultry products. Some, such as pre-stuffed whole birds, must be cooked from the frozen state to ensure a safely cooked product.

Freezer Storage Chart (0 °F) Food	**Freezer**
Meats, Fresh	
Beef or venison, roast, steaks	6-12 months
Chicken or turkey, pieces	9-12 months
Chicken or turkey, whole	1 year
Ducks, geese, game birds	6 months
Ground meat or stew, giblets	3-4 months
Lamb, roast or chops	6-9 months
Pork, roasts or chops	4-6 months
Pre-stuffed chops or chicken breast	*
Sausage	1-2 months
Variety meats: heart, liver, etc.	3-4 months
Meats, Cooked	
Smoked breakfast sausage	1-2 months
Ham, hotdogs, luncheon meats	1-2 months
Cooked, leftover meat, gravy	2-3 months
Cooked, leftover poultry	4-6 months
Leftover chicken nuggets	1-3 months

Seafood, Fresh	
Fresh lean fish: cod, flounder, trout,haddock, halibut, pollack, perch	4-6 months
Fresh fatty fish: Mullet, salmon, swordfish, mackerel, bluefish, tuna smelt	2-3 months
Shellfish	3 months
Cooked Fish	
Fish sticks, commercial	18 months
Breaded shrimp, commercial	1 year
Breaded scallops, commercial	16 months
Cooked pieces	3 months

Notes:

*Not recommended due to safety issues.

Freezing Prepared Foods

Preparing food ahead of time and freezing it can save time, energy and money.

Frozen cooked foods also add variety to your menu, offer quick meals for unexpected company and provide nutritious choices for busy days. When you are preparing a main dish, it takes only a little more effort and time to make enough for several meals.

You can freeze all of the prepared food in meal-size packages, or serve part of the food immediately and freeze the rest.

It is more economical to make your own supply of prepared dishes than to purchase commercially prepared foods.

Consider freezing:

- Leftovers that cannot be used immediately.
- Foods that take a long time to prepare.
- Seasonal fruits and vegetables.
- Foods you can prepare in quantity.
- Foods that still taste good after storage.

Hints for Freezing

- Select only fresh, high-quality ingredients because freezing does not improve quality.
- Slightly undercook prepared foods. They will finish cooking when reheated.
- Cool foods quickly before packaging. Place the pan of food in a

large pan of ice water, crushed ice or ice cubes. Stirring will help cool the food faster. Use a fan to cool foods that cannot be stirred.

- Freeze food promptly as soon as it is cooled to room temperature.
- Put no more unfrozen food in the freezer than will freeze within 24 hours. Usually this is 2 or 3 pounds per cubic foot of freezer capacity. Stack the food after it is frozen.
- Plan to use frozen prepared foods within a short time. Keep using foods from the freezer and replenish with fresh stock. This makes greater use of freezer space, lowers the cost per pound of food stores, and keeps your store of food fresh.
- The temperature of the freezer should not go above 0°F. Fluctuating temperatures and temperatures above 0°F reduce quality.
- Foods that do not freeze well include mayonnaise, cream puddings and fillings, custard, gelatin salads, cheese, the whites of hard-cooked eggs and uncooked egg yolks, and gravies made with wheat flour.
- Since spices may change flavour over long storage, add just before serving.

Hints for Packaging

- Use freezer containers or wrappings of moisture-and vapour-resistant material.
- Pack food compactly into the container to reduce air in the package. Allow headroom for expansion as food freezes.
- In quart containers, the food may be separated into two or three layers by a double thickness of a water-resistant wrap.
- Choose containers by the number of servings you will want. Quart containers hold four to six servings; pints, two to three.
- Use only containers with wide top openings. Food can then be removed without thawing.
- Freeze the prepared foods in your favourite casserole dish. The food can be removed after freezing, wrapped and returned to the freezer for storage.
- For microwave reheating, use microwave-safe plastic wrap for wrapping small amounts of breads. Casserole containers should be approved for micro-wave use.
- Label and date all packages, and keep an inventory of all frozen food.

Problem Foods

- Unbaked biscuits will be smaller and less tender. Unbaked muffins are likely to have poor texture.
- Custards and cream pies soak into the crust.
- Meringue on meringue pies toughens and sticks to the wrapper.
- Milk sauces sometimes curdle and separate. Stirring while reheating helps keep them smooth. Using waxy rice flour or waxy corn flour as the thickener also helps.
- Gravy tends to separate and curdle when thawed. It is better to freeze broth and make gravy just before serving. Or use waxy rice flour or waxy corn flour as the thickener.
- Cooked, creamed vegetables tend to lose flavour rapidly and should only be stored for a few days. Cook vegetables. Cool quickly. Add sauce. Package. Leave headspace.
- Lettuce, other greens and raw tomatoes lose crispness and become soggy.
- Ham and other cured meats may lose colour when frozen and become rancid more quickly than other meats.

Freezing Fruits, Step By Step

Harvested, fresh fruits continue to undergo chemical changes that can cause spoilage and deterioration of the product.

This is why these products should be frozen as soon after harvest as possible and at their peak degree of ripeness. Many fresh fruits, including peaches, apples, pears, plums, nectarines and sweet cherries will darken if their cut surfaces are exposed to air. This darkening will continue in the freezer if an anti-darkening treatment is not used. Because fruits are usually served raw, they are not blanched like vegetables. Ascorbic acid may be added to packing liquids or sprinkled on cut fruit surfaces to prevent darkening. Ascorbic acid is available at drug stores and is the major active ingredient in commercial anti-darkening preparations such as Fruit Fresh™.

Other methods to control browning include soaking the fruit in dilute vinegar solutions or coating the fruit with sugar and lemon juice. However, these latter methods do not prevent browning as effectively as treatment with ascorbic acid. Apples and rhubarb may be cooked to prevent browning. Rancid oxidative flavours may develop through contact of the frozen product with air. This problem can be prevented by using a wrapping material that does not permit air to pass into the product and by removing as much air as possible from the freezer bag or container before freezing.

Step-By-Step Instructions

Pack Fruit for Freezer

Most fruits have better texture if they are packed in sugar or syrup, however sugar is not required for preservation. Some fruits, including blueberries, cranberries, currants, figs, gooseberries, raspberries, rhubarb and steamed apples yield high-quality packs without sugar.

Your selection of the way to pack the fruit will also depend on the intended use. Fruits packed in syrup are generally best for uncooked desserts; those packed in dry sugar or unsweetened are best for most cooking purposes because there is less liquid in the product.

Artificial sweeteners—like saccharin or aspartame—may give fruits a sweet flavour but do not furnish the beneficial effects of sugar, such as colour protection and thickness of syrup.

Artificial sweeteners can be added just before serving:

- *Syrup Pack*: Pack fruit into freezer container, leaving ½-to 1-inch headspace. Cover fruit with syrup. To keep fruit under syrup, place a small piece of crumpled waxed paper on top and press fruit down into liquid.
- *Sugar Pack*: Cut fruit into a bowl or shallow pan. Sprinkle the sugar over the fruit. Mix gently with a large, long-handled spoon until juice is drawn out and sugar is dissolved. Put fruit and juice in containers, leaving ½-to 1-inch headspace.
- *Dry Pack:* The dry pack is good for small whole fruits such as berries, which give a good quality product without sugar. Simply pack the fruit into a container, seal and freeze. Or freeze individual berries or melon balls on waxed paper-lined trays and package them in containers as soon as they are frozen, to prevent freezer burn.
- *Other Unsweetened Packs:* Pack prepared fruit into containers and cover with water containing ascorbic acid. Or pack crushed or sliced fruit in its own juice, mixing in 1 teaspoon of ascorbic acid per quart of fruit, if necessary. Unsweetened packs will yield a product without the plump texture and good colour of those packed with sugar.

Seal Containers

- To remove air from plastic containers, seal container, then lift one corner of the lid pressing down the center of the lid. Reseal.
- If you use freezer bags for a liquid pack, fill them to within 3 inches of the top. If not using zipper lock freezer bags, seal by twisting the top of the bag where the food ends. Double the twisted top

back and close it with a rubber band or metal tie about ½ inch from the food to allow food to expand.

- If you use freezer bags for dry pack, fill them to within 3 inches of the top. Lower filled bag into a container of cold water, but do not allow water to get into container; this will remove much of the air from the container. Seal as explained above for a liquid pack in freezer bags.

Label Each Container with Name of Product and Date Package

Use the fruit within eight to twelve months for best quality. Unsweetened fruits lose quality faster than fruits packed in sugar or syrups.

Drying Foods

Food drying is one of the oldest methods of preserving food for later use. It can either be an alternative to canning and freezing or compliment these methods. Drying food is simple, safe and easy to learn. With modern food dehydrators fruit leathers, banana chips, pumpkin seeds and beef jerky can all be dried year-round at home. Dried foods are ideal for backpacking and camping. They are lightweight, take up little space and do not require refrigeration.

How Drying Preserves Food

Drying removes the moisture from the food so that bacteria, yeasts and molds cannot grow and spoil the food. It also slows down the action of enzymes but does not inactivate them.

When the food is ready for use, the water is added back and the food returns to its original shape. Foods can be dried in the sun, in an oven or in a food dehydrator by using the right combination of warm temperatures, low humidity and air current.

The optimum temperature for drying food is 140 °F. If higher temperatures are used, the food will cook instead of drying. When the food cooks on the outside and the moisture cannot escape, "case hardening" can occur. The food will eventually mold. Thus, the drying process should never be hurried by raising the drying temperature. Low humidity aids the drying process. Food contains a lot of water. To dry food, the water must move from the food to the surrounding air. If the surrounding air is humid, then drying will be slowed down. Increasing the air current speeds up drying by moving the surrounding moist air away from the food. To speed the drying time, increase the air flow. Most foods can be dried indoors using modern food dehydrators, counter-top convection ovens or conventional ovens. Microwave ovens are recommended only for drying herbs, because there is no way to create enough air flow to dry denser foods.

Sun Drying

Vegetables and meats are not recommended for out-of-doors drying. Vegetables are low in sugar and acid, which increases the risks for food spoilage.

Meats are high in protein, making them ideal for microbial growth when heat and humidity can't be controlled. It is best to dry meats and vegetables indoors using controlled conditions of an oven or food dehydrator.

The high sugar and acid content of fruits make them safe to dry out-of-doors when conditions are favourable for drying. It takes several days to dry foods out-of-doors, and because the weather is uncontrollable, this method can be risky. Hot, breezy days with humidity below 60 per cent are best, but usually these ideal conditions are not available in South Carolina when the fruit ripens, so that alternative methods are needed.

A minimum temperature of 85 °F is needed with higher temperatures being better. Fruits dried out-of-doors must be covered or brought under shelter at night. The cool night air condenses and could add moisture back to the food, thus slowing down the drying process.

Equipment

Racks or screens placed on blocks allow for better air movement around the food. Because the ground may be moist, it is best to place the racks or screens on a concrete driveway or if possible over a sheet of aluminum or tin. The reflection of the sun on the metal increases the drying temperature. Screens need to be safe for contact with food. The best screens are stainless steel, teflon-coated fibreglass and plastic. Avoid screens made from "hardware cloth." This is galvanized metal cloth that is coated with cadmium or zinc. These metals can oxidize, leaving harmful residues on the food. Also avoid copper and aluminum screening. Copper destroys vitamin C and increases oxidation. Aluminum tends to discolour and corrode. Because birds and insects are attracted to dried fruits, two screens are best for drying food. One screen acts as a shelf and the other as a protective cover. Cheesecloth could also be used to cover the food.

Solar Drying

Recent efforts to improve sun drying have led to solar drying. Solar drying uses the sun as the heat source, but a specially designed dehydrator increases the temperature and air current to speed up the drying time. Shorter drying times reduce the risk of food spoilage or molding.

Vine Drying

Another method of drying outdoors is vine drying. To dry beans leave bean pods on the vine in the garden until the beans inside rattle. When the vines and pods are dry and shriveled, pick the beans and shell them. No pretreatment is necessary. If beans are still moist, the drying process is not complete and the beans will mold if not more thoroughly dried.

If needed, drying can be completed in the sun, oven or a dehydrator:

- *Pasteurization*: Sun-dried fruits and vine-dried beans need treatment to kill insects and their eggs.
- *Freezer Method:* Seal the food in freezer-type plastic bags. Place the bags in a freezer set at 0 °F or below and leave them at least 48 hours.
- *Oven Method:* Place the food in a single layer on a tray or in a shallow pan. Place in an oven preheated to 160 °F for 30 minutes.

Food Dehydrators

A food dehydrator is a small electrical appliance for drying foods indoors. A food dehydrator has an electric element for heat and a fan and vents for air circulation. Dehydrators are efficiently designed to dry foods quickly at 140 °F. Food dehydrators are available from department stores, mail-order catalogs, natural food stores, and seed or garden supply catalogs. Costs vary from $50 to $350 or above depending on features.

Some models are expandable and additional trays can be purchased later. Twelve square feet of drying space dries about a half-bushel of produce. The major disadvantage of a dehydrator is its limited capacity.

Dehydrator features to look for:

- Double wall construction of metal or high-grade plastic. Wood is not recommended, because it is a fire hazard and is difficult to clean.
- Enclosed heating elements.
- Counter-top design.
- An enclosed thermostat from 85 to 160 °F and a dial for regulating temperature.
- A fan or blower.
- Four to 10 open mesh trays made of sturdy lightweight plastic for easy washing.
- A timer to turn the dehydrator off and prevent scorching if the drying time is completed during the night.
- UL seal of approval, a one-year guarantee and convenient service.

Types of Dehydrators

There are two basic designs for dehydrators. In horizontal air flow units, the heating element and fan are located on the side, whereas the vertical air flow dehydrators have the heating element and fan located at the base.

The major advantages of horizontal flow are:

- It reduces flavour mixture so several different foods can be dried at one time; all trays receive equal heat penetration;
- Juices or liquids do not drip down into the heating element.

Oven Drying

Everyone who has an oven has a food dehydrator. By combining the factors of heat, low humidity and air current, an oven can be used as a dehydrator.

An oven is ideal for occasional drying of meat jerkies, fruit leathers, banana chips or for preserving excess produce like celery or mushrooms. Because the oven may also be needed for everyday cooking, it may not be satisfactory for preserving abundant garden produce.

Oven drying is slower than dehydrators because it does not have a built-in fan for the air movement. It takes twice as long to dry food in an oven than in a dehydrator, and it uses more energy.

To Use your Oven

First, check your dial and see if it has a reading as low as 140 °F. If your oven does not go this low, then your food will cook instead of dry. For air circulation, leave the oven door propped open 2 to 6 inches.

Circulation can be improved by placing a fan outside the oven near the door. CAUTION: This is not a safe practice for a home with small children. Because the door is left open, the temperature will vary.

An oven thermometer placed near the food gives an accurate reading. Adjust the temperature dial to achieve the needed 140 °F. Trays should be narrow enough to clear the sides of the oven and should be 3 to 4 inches shorter than the oven from front to back. Cake cooling racks placed on top of cookie sheets work well for some foods. The oven racks, holding the trays, should be 2 to 3 inches apart for air circulation.

Room Drying

This method of drying differs from sun drying since it takes place indoors in a well-ventilated attic, room, car, camper or screened-in-porch. Herbs, hot peppers, nuts in the shell and partially sun-dried fruits are the most common air-dried items. Herbs and peppers can be strung on a string or tied in bundles and suspended from overhead racks in the air until dry.

Enclosing them in paper bags, with openings for air circulation, protects them from dust, loose insulation and other pollutants. Nuts are spread on papers, a single layer thick. Partially sundried fruits should be left on their drying trays.

Dehydrofreezing

Dehydrofreezing is a new method of food preservation that combines the techniques of drying and freezing. Fruits dried at home normally have had 80 per cent of their moisture removed; vegetables, 90 per cent.

However, by removing only 70 per cent of the moisture and storing the fruit or vegetable in the freezer, a tastier product results. The low temperature of the freezer inhibits microbial growth.

Also, the food takes up less room in the freezer. Dehydrofrozen fruits and vegetables have good flavour and colour. They reconstitute in about one-half the time it takes for traditionally dried foods. Dehydrofreezing is not freeze-drying.

Freeze-drying is a commercial technique that forms a vacuum while the food is freezing. Freeze-drying is a costly process that can't be done in the home.

Packaging and Storing Dried Foods

Dried foods are susceptible to insect contamination and moisture reabsorption and must be properly packaged and stored immediately. First, cool completely.

Warm food causes sweating which could provide enough moisture for mold to grow. Pack foods into clean, dry insect-proof containers as tightly as possible without crushing. Glass jars, metal cans or boxes with tightly fitted lids or moisture-vapour resistant freezer cartons make good containers for storing dried foods. Heavy-duty plastic bags are acceptable but are not insect-and rodent-proof.

Pack food in amounts that will be used in a recipe. Every time a package is re-opened, the food is exposed to air and moisture that lower the quality of the food.

Fruit that has been sulfured should not touch metal. Place the fruit in a plastic bag before storing it in a metal can. Sulfur fumes will react with the metal and cause colour changes in the fruit. Dried foods should be stored in cool, dry, dark areas. Recommended storage times for dried foods range from four months to one year.

Because food quality is affected by heat, the storage temperature helps determine the length of storage; the higher the temperature, the shorter the storage time. Most dried fruits can be stored for one year at 60°F, six months at 80°F. Vegetables have about half the shelf-life of fruits. Foods that are packaged seemingly "bone-dry" can spoil if moisture is reabsorbed during storage. Check dried foods frequently during storage to see if they are still dry. Glass containers are excellent for storage because any moisture that collects on the inside can be seen easily. Foods affected by moisture, but not spoiled, should be used immediately or redried and repackaged. Moldy foods should be discarded.

Bibliography

A S Karwa; M K Rai and H B Singh: *Handbook of Techniques in Microbiology : A Laboratory Guide to Microbes*, Scientific Publication, Jaipur, 2008.

A.K. Shrivastava : *Agriculture and Food*, APH Publication, Delhi, 2004.

Andrew L. Winton and Kate Barber Winton : *A Handbook of Structure and Composition of Foods*, Agrobios Publication, Delhi, 2013.

Andrew L. Winton and Kate Barber Winton:*Milk and Milk Products*, Agrobios Publication, Jaipur, 2002.

Anubhav Dwivedi : *A Fragrance from Food Production World*, Axis Publication, Delhi, 2010.

Arnold C. Long : *Advances in Seafood Biochemistry: Composition and Quality*, Cyber Tech Publication, Delhi, 2009.

Aysha Aziz Faridi:*Dairy Microbiology*, Random Publications, Delhi, 2012.

B. K. Mishra:*Dairy and Food Processing Industry : Recent Trends (2 Vols-Set)*, Biotech Books, New Delhi, 2014.

Boria Majumdar : *Cooking on the Run : An Average Indian Man's Encounters with Food*, Harper Collins, Delhi, 2012.

Bruno Dorin and Frederic Landy : *Agriculture and Food in India : A Half-Century Review*, Manohar Publication, Delhi, 2009.

Gilles Dussap : *Current Topics on Bioprocesses in Food Industry*, Asiatech Publication, Delhi, 2006.

H. Meliczek and C. Kaetsch : *Crop Science and Land Use for Food and Bioenergy*, Agrobios Publication, Delhi, 2010.

M K Rao:*Food and Dairy Microbiology*, Manglam Publication, Delhi, 2007.

M. Singh, D.K. Sharma and U.K. Mishra:*Organic Dairy Farming*, Satish Serial Publication, Delhi, 2011.

Mangala Kango:*Milk and Milk Products*, RBSA Publication, Jaipur, 2006.

Manoj Negi: *Career in Dairy Farming*, Abhishek Publication, Delhi, 2010.

Mohammad Raziuddin and Ashok Hembade: *Milk and Milk Products Technology*, Jaya Publishing House, Delhi, 2013.

P.C. Trivedi:*Microbes : Applications and Effects*, Aavishkar Publication, Delhi, 2009.

Piyush Sharma and Prem Ram : *A Textbook of Food and Beverage Management*, Dominant Publication, Delhi, 2013.

Rajiv K Sinha and Rohit Sinha: *Environmental Biotechnology : Role of Plants, Microbes and Earthworms in Environmental Management and Sustainable Development*, Aavishkar Publication, Delhi, 2008.

Ramniwas Sharma : *Agriculture and Food Safety*, Biotech Books, Delhi, 2006.

S. Sundara Rajan: *Genetics and Physiology of Microbes*, Anmol Publication, 2003.

S.K. Soni: *Microbes : A Source of Energy for 21 Century*, New India Publication, Delhi, 2007.

Urvashi Nandal : *A Handbook of Foods and Nutritional Biochemistry*, Agrobios Publication, Delhi, 2013.

Vanete Thomaz Soccol and Christian Larroche : *Advances in Bioprocesses in Food Industry*, Asiatech Publication, Delhi, 2011.

Vanete Thomaz Soccol and Christian Larroche : *Advances in Bioprocesses in Food Industry*, Asiatech Publication, Delhi, 2011.

Varnam: *Milk and Milk Products: Technology, Chemistry, and Microbiology*, Springer Publication, London, 2003.

W. Merback and A. Veha : *Crop Science and Technology for Food Security Bioenergy and Sustainability*, L. Bona, J. Pauk, , Agrobios Publication, Delhi, 2012.

W.L. Slatter, T. Krishtofferson and D.A. Seiberling: *Manual of Dairy Industry : Training and Development of Personnel for the Dairy Industry*, Asiatic Publication, Delhi, 2012.

Index